What people are say

Dan Hull's background and recognized experience as a practicing engineer and national leader in technical education reform uniquely qualifies him to understand the need and create this solution for producing more qualified technicians.

Betsy Brand, Executive Director of American Youth Policy Forum,
Former Assistant U.S. Secretary of Education

Career Pathways for STEM Technicians is a timely effort. The Advanced Technological Education (ATE) program at the National Science Foundation (NSF) has a wealth of expertise and materials that support career pathways and technician education programs. It is critical that the greater educational community work with industry to develop and implement programs that integrate rigorous STEM courses with industry recognized skills and competences. Dan Hull and his contributors present a compelling case for expanding existing efforts to meet the critical need for a qualified STEM technical workforce in the United States.

Celeste Carter, Lead Program Director ATE Program
Division of Undergraduate Education, National Science Foundation

---

This book clearly outlines the need and presents information required for an educational reform that will prepare more young people for meaningful and exciting careers in numerous fields that employ lasers and optics technology. Industry leaders and the small business community must be proactive to assure that these reforms are implemented in order to have access to the qualified technicians required to meet the workforce demands of the future.

Larry Dosser, President and CEO, Mound Laser and Photonics Center, OH

CPST should not be undertaken as a boutique program for a talented few, but as a means of enhancing success for all students.

Jim Jacobs, President Macomb Community College, MI

Career Pathways for STEM Technicians takes a comprehensive approach to the why, what and how of improving the education and "adulthood-prep" experience for the vast majority of our students. It looks at education within a society-wide context and the context of preparing the whole child.

Willard R. Daggett, CEO,
The International Center for Leadership in Education

Technical students possess superior spatial learning abilities than non-technical students at two-year colleges in the United States.

Darrell Hull, University of North Texas

Dan Hull and his colleagues have provided high schools, community colleges, and employers with practical recommendations for developing CPST pathway models that work. Dan's lifelong commitment to Career Technical Education continues to inspire and guide STEM administrators, instructors, and policy makers around the country and the globe!

David D. Gatewood, Dean of CTE
Irvine Valley College, CA

Today we are faced with a paradox, high tech companies are seriously seeking qualified workers and yet program enrollments in career and technology education programs are not coming close to satisfying industry's workforce needs. ***Career Pathways for STEM Technicians* calls us to action and gives us promising practices to use to address this gap.**

Michael Lesiecki, Director, MATEC,
Maricopa Community Colleges, AZ

The technical workforce we need for national security and economic prosperity is like a three-legged stool: scientists, engineers and technicians. The "shortest leg" of this stool is technicians. They have the unique role as *geniuses of the labs* and *masters of the equipment.*

Dan Hull

CAREER PATHWAYS
FOR
STEM TECHNICIANS

Dan M. Hull   July 2012

This document is based on work supported by the National Science Foundation under Grant No. NSF/DUE 1203500. Any opinions, findings, and conclusions or recommendations expressed in this document are those of the authors and do not necessarily reflect the views of the National Science Foundation.

Daniel Hull, PI
324B Kelly Dr
Waco TX 76710
(254) 751-9000
hull@op-tec.org

ISBN 978-0-9858006-1-1

CONTENTS

Foreword

In my 2011 *STEM* report,[i] I cite the following findings:

- Science, Technology, Engineering, and Mathematics (STEM) occupations are critical to our continued economic competitiveness because of their direct ties to innovation, economic growth, and productivity.

- The STEM supply problem goes beyond the need for more professional scientists, engineers, and mathematicians. We also need more qualified technicians and skilled STEM workers in Advanced Manufacturing, Utilities and Transportation, Mining, and other technology-driven industries. What's more, people within these occupations who use STEM competencies most intensely are earning significantly more than those who are not.

- Our K–12 education system produces enough talent in math and science to fill our need for traditional STEM workers, but more than 75 percent of these students do not enter STEM majors in college.

The vast majority of STEM jobs require some form of postsecondary education or training.

- By 2018, roughly 35 percent of the STEM workforce will be composed of those with sub-baccalaureate training, including:
  - ✓ 1 million associate degrees,
  - ✓ 745,000 certificates, and
  - ✓ 760,000 industry-based certifications.

We find that STEM wages are high and have kept up with wages as a whole over the last thirty years.

- Two-thirds (66%) of STEM workers with an associate degree make more than the average for workers without an associate degree.

- Wages for engineers and engineering technicians have grown 18 percent since the early 1980s. The average salary for engineers and engineering technicians ($78,000) is higher than salaries for all other STEM occupations.

Going forward, we will need more workers with STEM competencies—but not necessarily traditional STEM workers in traditional STEM jobs.

- Concern for the supply of the highest-performing STEM workers tends to point toward strategies targeted at relatively small portions of American students among our top science and math performers. However, these elite workers are not the entirety of the STEM workforce.

- While many remain focused on a small cadre of elite STEM workers, more than a third of all jobs in STEM through 2018 will be for those with less than a bachelor's degree. There is increasing demand for STEM talent at the sub-baccalaureate level, and our education system has, thus far, not adequately produced these workers. Going forward, our Career and Technical Education system will need a stronger STEM curriculum at the high school and sub-baccalaureate level that is more tightly linked with competencies necessary for STEM jobs.

Dan Hull and the other collaborators in *Career Pathways for STEM Technicians* present a practical, workable strategy to build on the resources and infrastructure that are already in place to strengthen our nation's technical workforce—and to provide rewarding career opportunities for a neglected group of deserving high school students. I strongly recommend your thoughtful consideration of this innovative initiative.

Anthony Carnevale, 2012

---

[i] Anthony P. Carnevale, Nicole Smith, and Michelle Melton, *STEM* (Washington, DC: Center on Education and the Workforce, Georgetown University, October 2011), http://cew.georgetown.edu/stem/ (accessed June 6, 2012).

Preface

This book presents **a "win-win" solution to two problems** that are facing our nation:

- *We don't have enough technicians to support continued technological innovation or to staff the organizations that could improve our country's economic position in the world.*
- *We don't have adequate educational opportunities for capable, struggling high school students who need—and deserve—an opportunity to enter a rewarding career.*

We have the **tools and institutions in place** to implement the proposed solution:

- *Associate-degree technical-education programs.* Our community and technical colleges offer these programs using curricula and teaching materials that have been specified by the industries and employers who want to hire technicians. They are taught by competent faculty members who have been trained in both the content and in the strategies that can help technical students succeed in their careers.
- *High school STEM programs and academies.* Specialized STEM programs have been formed over the past decade to interest, attract, and cultivate students to enter postsecondary education in preparation for careers in engineering, science, and technology.

The dilemma we face is this:

- Most of the colleges offering technician education in the new and emerging technologies *do not have an adequate number of capable students* enrolling in and completing their associate-degree programs.
- Most high school STEM programs are *not attracting and serving the students who have the greatest potential to become the technicians we need.*

The solution this book presents is not prohibitively expensive or excessively time consuming; neither does it require that existing institutions and programs sacrifice the goals or quality of the programs that they have already put in place. But the solution does require some changes:

- *Recognition*—by educators and parents—that the high school students who are our potential technicians are not receiving the career guidance and education that they need.
- *Engagement* in a process that will identify students who may be interested in becoming technicians, and encouragement for those students to enter a career pathway that will lead them through an associate-degree program and on to a career as a technician.
- A *commitment* to providing an alternative career pathway in STEM high schools that will serve these students through dual-credit and articulated secondary/postsecondary coursework.

I left engineering and management of emerging technologies after thirteen years because I wanted to assure that in future years we would have an adequate supply of well-prepared technicians. I then spent over thirty years in technician-education research and development because of my concern and passion for capable high school students who typically are not in the top quartile of academic achievers, mostly because they are applied learners—not abstract learners. These students are potentially our future technicians, and they deserve the respect, guidance, and educational opportunity to become successful in technical careers.

Creating this book about career pathways for STEM technicians has been a labor of love. In this endeavor I have been inspired, blessed, and supported by colleagues, coauthors, family, an incredible editor (Kelly Besecke) and a talented designer (Rachel Haferkamp). Thank you all!

Daniel Hull
Registered Professional Engineer   2012

Contributors

JUNE ST. CLAIR ATKINSON, EdD, is the State Superintendent of Public Instruction in North Carolina and has served in that role since 2005. During her career in education, she has been a public school teacher, community college instructor, and state administrator. As a member of the North Carolina Department of Public Instruction, she has been the chief state school officer, Director of Career-Technical Education, and Director of Instructional Services. Dr. Atkinson has also served as President of the National Business Education Association and the National Association of State Directors for Career-Technical Education Consortium.

ANN BEHELER, PhD, is the Principal Investigator for the National Science Foundation Advanced Technological Education (NSF/ATE) IT Convergence Technology Center. She has worked in the information technology industry for over thirty years. She is currently responsible for emerging technology grants at Collin College in Texas. Before that, she was Vice President of Academic Affairs for Porterville College in California. She partners colleges with the IT industry, using a streamlined process to identify the knowledge, skills, and abilities (KSAs) for future IT convergence technicians, and then working with faculty members to design appropriate curriculum to educate and train technicians to earn various certificates and degrees in IT.

GEORGE BOGGS, PhD, is President and CEO Emeritus of the American Association of Community Colleges (AACC), where he served for over ten years as a leading postsecondary education advocate and spokesperson. Dr. Boggs served as a faculty member and administrator at Butte College in California and, for more than fifteen years, as the Superintendent/President of Palomar College in California. He serves as a member of the Board on Science Education of the National Academy of Sciences and has served on several U.S. National Science Board and Foundation panels, commissions, and committees.

ANTHONY P. CARNEVALE, PhD, is Director of the Georgetown University Center on Education and the Workforce. Dr. Carnevale has served as Vice President for Public Leadership at the Educational Testing Service (ETS). While at ETS, he was appointed by President George W. Bush to serve on the White House Commission on Technology and Adult Education. Before joining ETS, Dr. Carnevale was Director of Human Resource and Employment Studies at the Committee for Economic Development (CED), the nation's oldest business-sponsored policy research organization. While at CED, he was appointed by President Bill Clinton to chair the National Commission on Employment Policy.

MEL COSSETTE, MEd, is the Executive Director and Principal Investigator for the National Science Foundation–funded National Resource Center for Material

Technology Education (MatEd), housed at Edmonds Community College in Lynnwood, WA. Mel has over twenty years of experience in manufacturing education and has developed technician education and training programs for industry and educational institutions. She serves on numerous committees and national boards and worked in various industries prior to holding administrative positions in the community and technical college system. Recent research focused on identifying successful practices in the recruitment of women to STEM fields.

WILLARD DAGGETT, EdD, Founder/Chairman of the International Center for Leadership in Education, is nationally recognized for his efforts in school improvement. He has advised governors, state education officials, districts, and schools across America on education reform. Dr. Daggett is the creator of the Rigor/Relevance Framework®, a conceptual model and practical blueprint for designing college- and career-ready curriculum and instruction for *all* students. Bill challenges educators and their stakeholders to embrace what is best in American education while making the changes needed to prepare students for careers, citizenship, and lifelong learning in the twenty-first century.

KATHY D'ANTONI, EdD, is the Assistant State Superintendent for the West Virginia School System for Career Technical, Adult, and Institutional Education. Her career in education includes teaching positions, serving as interim president of a community and technical college, working with the Tech Prep initiative at Marshall University, and serving as West Virginia's State Director for Tech Prep. Dr. D'Antoni has worked extensively with curriculum alignment and development projects and is the past president of the National Association for Tech Prep Leadership. She has authored numerous articles on effective transitions from public schools to higher education.

J. D. HOYE is President of the National Academy Foundation, a network of five hundred academies that open doors for students to professional careers. For thirty-five years, she has worked to better position young people to pursue their academic and career goals, becoming a national leader in educator/employer partnerships. During the Clinton administration, she was appointed to lead the Office of School-to-Work, spurring nationwide progress in college and career readiness. She was previously Associate Superintendent of the Office of Professional/Technical Education for the Oregon Department of Education and Office of Community Colleges.

KATHERINE HUGHES, PhD, is the Assistant Director for Work and Education Reform Research at both the Institute on Education and the Economy and the Community College Research Center at Teachers College, Columbia University. Dr. Hughes specializes in research on the transition from high school to college and careers, including projects addressing secondary-postsecondary partnerships, career-technical education, dual enrollment, state policies that facilitate transitions and pathways, incoming community college student assessment and placement, student retention, and work-based learning, among other topics.

DANIEL HULL is a Registered Professional Engineer with thirteen years of practice in the laser field and over thirty years of experience leading education reform efforts across the United States and internationally. In 2006, he established the NSF/ATE National Center for Optics and Photonics Education (OP-TEC); he continues to serve as Principal Investigator and Executive Director of OP-TEC.

In 1979, Dan established CORD (Center for Occupational Research and Development); he served as CORD's CEO until 2006. He also established the National Coalition for Advanced Technology Centers and the National Tech Prep Network (now the National Career Pathways Network). He conceived and collaborated in the creation of the NSF center-sponsored, annual HI-TEC Conference. Dan is the author of four books on Tech Prep and contextual teaching. His most recent books are *Career Pathways: Education With a Purpose* (2005)*, and Adult Career Pathways* (2007). He frequently collaborates with other subject-matter experts to coauthor these books. Dan has been elected Senior Member of SPIE (the international society for optics and photonics), the Optical Society of America, and the Laser Institute of America.

DARRELL M. HULL, PhD, is Assistant Professor of Research, Measurement, and Statistics in the Department of Educational Psychology at the University of North Texas. He completed an AAS degree in laser/electro-optics and was a technician before continuing his education, which eventually led to a PhD in Educational Psychology. In 1993, he developed the first edition of the National Skill Standards for Photonics Technicians, and he currently serves as Internal Evaluator for an NSF/ATE National Center. His research interests include the study of individual differences in cognitive ability and personality traits, and his work has included a recent study funded by the NSF/ATE targeted research program. In addition, he has conducted several experimental and quasi-experimental studies to examine effects of educational and training programs in the Caribbean for the Inter-American Development Bank.

JAMES JACOBS, PhD, has been the President of Macomb Community College since 2008. Prior to his appointment, he concurrently served as Director of the Center for Workforce Development and Policy at Macomb and as Associate Director of the Community College Research Center (CCRC) at Columbia University's Teachers College. Jim has more than forty years of experience at Macomb; he has taught social science, political science, and economics, and has specialized in community college workforce development and worker retraining. He has served as President of the National Council for Workforce Education. Jim has conducted research, developed technical programs, and consulted on workforce development and community college issues at the national, state, and local levels.

ELAINE A. JOHNSON, PhD, is the Executive Director of Bio-Link, a National Science Foundation Advanced Technological Education National Center for Biotechnology and Life Sciences based at City College of San Francisco. Elaine specializes in creating partnerships between educational institutions and industry. She promotes articulation in the effort to create career pathways. Elaine is nationally recognized as an innovator and leader in education for careers in biotechnology. In

2010, she received the John Blackburn Exemplary Models Award from the American Association for University Administrators.

GREG KEPNER, MEd, is Department Chair for Advanced Manufacturing Technology programs at Indian Hills Community College. He has administrative responsibility for the leadership of the manufacturing technology programs at IHCC. He has served as Industrial Technology Coordinator and has taught automation, robotics, and electronics. He developed an early-college program in which high school students earn postsecondary credits toward an AAS degree in Lasers, Robotics, or Electronics Engineering. He has worked as a Senior Field Service Engineer in semiconductor manufacturing. He is currently serving on the boards of the Iowa Association of Career and Technical Education and the Iowa Industrial Technology Education Association.

DEB M. NEWBERRY is Chair of NanoScience Technology at Dakota County Technical College and the Director/Principal Investigator for Nano-Link, an NSF/ATE Regional Center for Nanotechnology Education. Ms. Newberry's educational background includes nuclear physics (MS), chemical engineering, and mechanical engineering. She worked for twenty-three years in radiation analysis research for satellite systems and as Executive Director of Space Programs at General Dynamics. She is the coauthor of *The Next Big Thing Is Really Small: How Nanotechnology Will Impact Your Business,* published in 2003. Ms. Newberry has created eight college nanotechnology courses, including topics such as nanoelectronics, nanomaterials, and nanobiotechnology. Ms. Newberry is a member of the Institute of Electrical and Electronics Engineers (IEEE), the American Chemical Society (ACS), the American Association for the Advancement of Science (AAAS), the American Society for Engineering Education (ASEE), the Association of Technology, Management, and Applied Engineering (ATMAE), and the Institute of Physics (IOP

L. ALLEN PHELPS, PhD, is a Senior Scientist and Professor Emeritus at the Wisconsin Center for Education Research in the School of Education at the University of Wisconsin–Madison. Dr. Phelps's research and university teaching have focused on high school restructuring and innovation, two-year college leadership, career development, and equity for special populations. Most recently, he edited a special issue of *New Directions for Community Colleges*, entitled *Advancing the Regional Role of Two-Year Colleges.* He currently directs an NSF-funded longitudinal research study to document the individual, educational, and environmental factors influencing both short-term and long-term student success in technical college settings and the workplace.

MATTHIAS PLEIL, PhD, is the Principal Investigator for the Southwest Center for Microsystems Education (SCME). A Research Associate Professor of Mechanical Engineering at the University of New Mexico and a faculty member at Central New Mexico Community College, Dr. Pleil teaches courses in microsystems, physics, and engineering. He has twelve years of experience in semiconductor manufacturing, where he worked with technicians as a Senior Engineer and Engineering Manager.

He received his PhD in physics from Texas Tech University, where he completed original research on time-resolved fluorescence spectroscopy.

PAT SCHWALLIE-GIDDIS, PhD, is a national leader in school counseling and a recognized expert in career development. She is an Associate Professor at George Washington University and serves as the chair of the Department of Counseling and Human Development. Her past positions include serving as an executive staff member for the Association for Career and Technical Education and the American Counseling Association, and as a teacher, counselor, district-level administrator, and State Program Director for Career Development in her native state of Florida. She is the immediate Past President of the National Career Development Association, and has coauthored three books on counseling and career development.

JILL SILER, EdD, was the Executive Director of Academic and Organizational Development for Lake Travis Independent School District in Travis County, Texas. She worked primarily with secondary campuses, facilitating curriculum development, assessment creation, and professional learning opportunities for middle school and high school teachers. In addition, Jill oversaw strategic planning, the Institutes of Study, enrollment tracking, and staffing projections. Jill structured the Lake Travis High School College and Career Readiness program into the following six career pathways, called Institutes of Learning: Advanced Sciences and Medicine; Veterinary and Agricultural Science; Business, Finance, and Marketing; Math, Engineering, and Architecture; Humanities, Technology, and Communication; and Fine Arts. She is currently the Deputy Superintendent of Lake Travis ISD.

GORDON F. SNYDER, JR. is Executive Director of the NSF National Center for Information and Communications Technologies at Springfield Technical Community College Springfield Technical Community in Massachusetts. He is the author of four engineering and engineering-technology textbooks and has over twenty years of consulting experience in the field of communications design. He serves on several technology boards and in 2001 was selected as one of the top fifteen technology faculty members in the United States by Microsoft and the American Association of Community Colleges. His popular technology-focused tweets and blogs are followed by thousands.

JOHN C. SOUDERS, PhD, is the Director of Curriculum at the National Center for Optics and Photonics Education (OP-TEC). In this position, he has the responsibility of developing innovative and flexible curriculum that provides two-year colleges multiple pathways for integrating photonics offerings into their programs. Dr. Souders is the principal author of two nationally recognized sets of skill standards. He also led the development of three nationally recognized high school context-based mathematics books. He was a professor of physics at the Air Force Academy and served as Vice President of Academics at Cedar Valley College in Dallas, Texas.

ELIZABETH J. TELES, PhD, retired from NSF in January 2009 as the Lead Program Director of the Advanced Technological Education (ATE) Program in the Division of Undergraduate Education (DUE) and a program director for mathematics. At NSF, she was the Liaison for Community Colleges. In 2008, she received the NSF Meritorious Service Award, the second-highest honorary award at the Foundation, for her work on behalf of community colleges and her leadership in the ATE program. She now consults on educational projects and works as an intermittent expert for NSF. Liz taught mathematics at Montgomery College in Maryland from 1969 to 1991.

SONIA WALLMAN, PhD, is Executive Director of the NBC[2], a National Science Foundation Advanced Technological Education Center for Biomanufacturing. Her interest in career pathways in biomanufacturing developed when she worked with Lonza and other regional biopharmaceutical manufacturers to develop a biomanufacturing course based on industry skill standards at Great Bay Community College in Portsmouth, NH. The NBC[2] has worked with its partners to develop laboratory manuals, textbooks, virtual modules, and an online ancillary support site (http://www.biomanufacturing.org) dedicated to education and training in biopharmaceutical manufacturing and its crossover industries such as biofuels and regenerative medicine.

Our Country's Technology Edge is Fading Quickly:
And the Problem Will Not Be Solved with Just More Engineers and Scientists

Dan Hull
Executive Director, OP-TEC

The history of our country's emergence, leadership, and prosperity has shown evidence of our values and superiority as we have worked to make our nation a land of opportunity.

- In exploring and developing our land and investing in our people, *we have shown vision, courage, determination, and a commitment to hard work.*

- By creating democracy, public education, and a transportation system that was second to none, *we built an environment that supported the personal pursuit of opportunities and rewards.*

- *As we shifted from agriculture to construction, manufacturing, and services, we mobilized our workforce, infrastructure, and economy.*

- As we used our creativity and innovation to advance electronics, automation, energy resources, aerospace, defense, medicine, and information technology, *we captured and benefited from technological innovation.*

But over the last thirty years, we have been slipping in our international dominance and leadership because we have lost some of the characteristics that made us great:

- Our communities and educational systems are not developing sufficient globally competitive "human capital." *Superiority in science, engineering, technology, and innovation are important keys to a globally competitive workforce.*

- Instead of investing in the future, we are living in—and borrowing on—the past. *This "borrowing trend" is not limited to economics; we're also continuing to rely on the technical and innovative skills and abilities of a generation of workers that is rapidly approaching retirement.*
- We are not providing "opportunities for all" to our youth: *we are trying to educate everyone the same way, for the same jobs, as if every student had the same limited range of abilities and interests.*

It is encouraging to note that after thirty years of ignoring the problem or floundering with inadequate solutions, we are finally beginning to recognize the real issues and how urgent it is that we address them. In recent years, the media and communities throughout our country have begun to take part in a national dialogue about our faltering technological advantage. We all hope that we are not too late—and that our society and culture will have the vision and strategies to address these issues, the openness to reinvent ourselves, and the perseverance to complete the tasks that will help us succeed.

What will our nation become if we fail to maintain our edge in technology and innovation? How can we avert this risk?

These questions were the focus of *Rising Above the Gathering Storm*, a 2007 report by the National Academies of Engineering and Sciences (NAE and NAS) and the Institute of Medicine.[1] In response to a bipartisan request from the U.S. Congress, the Academies formed a knowledgeable, experienced, and prestigious committee that directed a staff to conduct a review of America's competitive position. In this very carefully worded, extensive report, these experts explained that we as a nation were in deep trouble. If our nation didn't turn things around, the report said, the United States was on track to become a second rate power—both economically and militarily—and future generations would not enjoy the prosperity and security that we have experienced over the last seventy years. *The Gathering Storm* presented four specific recommendations and supporting action plans:

1. Move the U.S. K–12 education system in science and mathematics to a leading position by global standards.
2. Double the federal investment in basic research in math, science, and engineering over the next seven years.
3. Encourage more U.S. citizens to pursue careers in math, science, and engineering.
4. Rebuild the competitive ecosystem through legislative reform.

Three years later, the National Academies revisited the situation to look for progress on their recommendations. They were disappointed to learn that little, if any, progress had been made. In 2010, the Academies released a new report, *Rising Above the Gathering Storm, Revisited: Rapidly Approaching Category 5.*[ii] The Committee found that although some progress had been made in research investment and legislative reform (recommendations 2 and 4), little or no progress had been made on K–12 science and math education or in the number of citizens pursuing STEM careers (recommendations 1 and 3). In fact, the authors noted that the quality of math and science education in the United States ranks forty-eighth in the world. They concluded that "the outlook for America to compete for quality jobs has still further deteriorated over the past five years."

So What Can We Do about It? Create New STEM Pathways in High Schools.
The Gathering Storm's first recommendation was to recruit more K–12 math and science teachers. We do need more and better-prepared math and science teachers, but the solutions that are being considered to address our teacher shortfall are expensive and take lots of time. Perhaps someday we will regard the teachers of our children with as much esteem and offer as many rewards as we do professional athletes and investment bankers. Meanwhile, we have other high school teachers who may be able to help improve the situation more immediately. These are the Career and Technology teachers who are providing basic foundations and experience in emerging technologies and stimulating students' interest in technology-centered careers.

Science, Technology, Engineering, and Mathematics (STEM) programs—often called "STEM academies"—have emerged rapidly in our nation's high schools. STEM programs are designed to interest students and prepare them to enter postsecondary education in STEM areas. An estimated 3,500 STEM programs are currently operating in America's high schools. Many of them are using a curriculum that is academically sound and technically valid and that provides the foundation necessary for graduates to enter and be successful in Bachelor of Science programs in science, engineering, information technology (IT), and mathematics.[iii]

As part of this curriculum, most STEM programs encourage students to begin taking higher-level math (pre-calculus and calculus) in their junior and senior years. But these abstract math courses end up actually deterring many potential STEM students from pursuing STEM careers. Calculus and pre-calculus courses become a *filter* that screens out students who are not interested in or successful at abstract math.

These students are unlikely to pursue baccalaureate studies in math, science, or engineering, but with different options for high school preparation, they may be interested in pursuing Associate of Applied Science (AAS) degrees in STEM fields. Students who fit this description are typically in the middle quartiles of achievement. They are capable of learning math and science at a useful level, but they are "applied learners," hands-on learners with great spatial abilities. They often ask "What's this good for?" and "How will I use this?" If they don't get a satisfactory answer (and "You're going to need it to pass the test" is not a satisfactory answer), they drop out—physically or mentally.[iv]

These students have the potential to enroll in our community and technical colleges and become the engineering, science, and IT technicians that U.S. employers so desperately need. But because most high school STEM curricula don't offer a career pathway to becoming a STEM technician, both students and employers are missing out.

The Three-Legged Stool of Our Technical Workforce

I was extremely fortunate to begin my engineering career during the emergence of lasers and electro-optics. For over thirteen years, I worked in corporate laboratories and field environments to support the Department of Defense, the Atomic Energy Commission (now the Department of Energy), and NASA. We were discovering new physical/optical phenomena; new laser materials and systems; new applications in defense and aerospace; and new challenges in the measurement of optical radiation, laser safety, and the fabrication and use of precision optics. I was a member of a technical team that consisted of scientists (physicists and spectroscopists), engineers (electrical and mechanical), and technicians. The scientists explored and characterized the theories; the engineers designed and tested the new equipment; and the technicians put the equipment together, made it operate, and kept it working. We all had our unique knowledge, talents, and abilities—and we were all needed on the team.

But we never had enough highly qualified technicians to complete our teams. Many of the technicians on our team had only a high school education; a few had gained some training and experience in the military. These undertrained technicians did not have the math and scientific background that would have helped them understand *why and how* our equipment worked. The few good technicians that had this background had earned AAS degrees from community and technical colleges. But there were simply not enough colleges educating these technicians.

Because we didn't have an adequate supply of well prepared technicians, I eventually left the field of engineering research and development and entered another career in technical education research and development. And nearly four decades later, we still don't have enough qualified technicians in our work force! Over the last two years, I have led discussions on this subject with focus groups of employers in a wide variety of fields from different regions of the country. In each of these meetings, employers have indicated that they're seeing at least as great a shortage of good technicians as they are of engineers and scientists.

What Are Technicians? Where Do They Come From?
First, let me explain what technicians are not. They aren't the "gofers" that only make coffee, run errands, or do manual labor in the shop.

The technicians that I worked with understood the technology and knew how to solve certain problems—but they were also the "geniuses of the labs" and the "masters of equipment." They were particularly good at hands-on technical tasks, and they had incredible spatial abilities. They could see how things should fit together, and they knew how to make equipment work.

The technicians I worked with could conduct experiments, run tests, and accurately record results. They understood the challenges that our team faced and made major contributions. Because these men and women were the graduates of two-year associate degree programs in our nation's community and technical colleges, they had mastered the field's fundamental mathematical and scientific principles, and this mastery supported their technical knowledge and skills.

Some technicians, instead of working in teams with engineers and scientists, work as technical support people in areas where hi-tech equipment is being used. They operate and maintain office computers, telecommunications systems, health care facilities, and energy-providing equipment. Sometimes technicians are called technologists or laboratory assistants.

Today, technicians are working with a wide variety of emerging technologies, such as photonics; nanotechnology; biotechnology; information and communication technology; advanced manufacturing; environmental monitoring; biomedical equipment; and nuclear, solar, wind, and other alternative energy fields, to name a few.

Associate degree and certificate education programs to prepare technicians are in place in many of our nation's more than 1,300 public community and technical colleges. Relevant curriculum materials and teaching strategies to support these programs are available and are continuously updated. Employers recognize the importance of technicians and readily support technical education programs in their local communities by advising on curriculum and labs, donating useful equipment, and providing adjunct faculty and internships.[v]

The Technician Shortage: Low Enrollment Due to Inadequate High School Pipeline

I am Director of the National Science Foundation Center of Excellence for Optics and Photonics Education (OP-TEC). The mission of our center is to encourage and support colleges that prepare photonics (lasers and optics) technicians, to assure that there is an adequate supply of these workers for our nation's industries. We determine how many photonics technicians are needed and the technical content that colleges should teach to students who are interested in becoming photonics technicians. We support community and technical colleges by designing curricula, developing teaching materials, and training faculty to teach new photonics courses.

In 2009, at the height of the recent depression, we commissioned the University of North Texas to conduct a survey of over four thousand U.S. photonics employers to determine their current and projected needs for technicians. This survey was updated in 2012. We also polled the thirty existing two-year photonics colleges to estimate the future supply of new technicians.

These surveys unearthed a huge disparity between supply and demand. The results of the recent study projected that employers will need to hire over eight hundred new photonics technicians each year. Meanwhile, only 750 students are currently enrolled in photonics programs, and fewer than three hundred students per year are completing their training and entering the work force.

These are good jobs. Photonics technicians with AAS degrees are earning starting salaries of $41,000–$56,000 per year. But because employers can't hire the qualified technicians they need in the United States, many are moving their operations off shore. Others are filling their technician positions with high school graduates and engineers, neither of which is a good fit. Employers prefer to hire educated technicians with AAS degrees, because most high school graduates don't have the necessary math and science background—and most engineers don't have the hands-on skills and spatial abilities.

OP-TEC is working with the existing photonics colleges to build their enrollment and reduce attrition to increase the supply of trained photonics technicians. We are also working with over thirty additional colleges that are planning to initiate new photonics programs. But "the dam is leaking quicker than we can plug the holes." Many programs are experiencing low enrollments; in the last five years, low enrollment has caused six programs to be closed. Fortunately, we have been able to help four of these programs reopen, but now others are at risk of closing due to low enrollments and tight budgets.

OP-TEC's challenge is not unique. The National Science Foundation's Advanced Technological Education (NSF/ATE) program is supporting almost forty national and regional centers, as well as many related projects, in nanotechnology, biotechnology, information technology, and other emerging fields. Nearly all these centers are experiencing similar low enrollments in the colleges they support.

So, Why Are Enrollments Low in AAS Programs that Prepare Technicians for These Rewarding and In-Demand Careers?

Public education in the United States has traditionally been perceived as a progressive, "upward mobility" ladder: students advance from K–6, to middle school, to high school, to higher education. Students have "climbed the ladder" as far as their interests, abilities, and resources have allowed them. Typically, the reward for climbing to the highest rungs has been a more rewarding career and a better-quality life style.

Prior to 1970, students could pursue rewarding careers with only a high school education, as long as they also acquired manual skills. In most cases, this opportunity no longer exists. Now, technical jobs require "head skills" as well as "hand skills," so nearly all new workers need some postsecondary education and training.

The problem is that our culture—and our approach to educational reform over the last twenty-five years—has fixated on *one single path* through higher education for *everyone.* High school curricula and teaching strategies are almost totally focused on preparing students to enter and succeed in four-year baccalaureate programs. Accordingly, high schools typically offer only one path toward higher education, and this path requires students to take abstract math courses in their junior or senior year.

This "one track" approach may appeal to the 20–25 percent of high school students who are abstract learners and achieve at high levels in pre-calculus and calculus, but it is excluding capable students in the middle quartiles who are more applied learners. These are the potential technician students who should be in courses that will articulate to community and technical colleges. *There is not a broad, well-established pipeline that encourages and prepares these students to enter postsecondary AAS degree programs to become the technicians we so desperately need.* As a consequence of this condition, not enough students apply for technician education at our community and technical colleges. Of the students who do apply, many are not prepared and are consequently at risk of dropping out in their first year. The results are obvious: not enough technician graduates.

Harvard Report Questions Value of "College for All"

This emphasis on preparing all students for baccalaureate programs is not unique to STEM programs; it has been the cornerstone of K–12 public school reform for the past twenty-five years.

But educational leaders have begun to recognize that the majority of students are not well served by this exclusive focus on four-year colleges. In a February 2011 report entitled *Pathways to Prosperity*, the Harvard Graduate School of Education criticized our public education system for concentrating too much on classroom-based academics and making four-year college the only goal.[vi] The report claims that the nation's education system has failed vast numbers of students who instead need solid preparation for careers that require degrees other than the baccalaureate.

Pathways to Prosperity supports a vision that would expand opportunity for all students, especially those who face the dimmest prospects because their education stops at high school. *Pathways* recommends that high schools offer sound college preparation; provide rigorous career-focused, real-world learning; and define clear career pathways from secondary school into certificate and associate degree programs.[vii] Rather than derailing some students from higher learning, this proposed system would actually open more pathways to higher education for more students. As the report observes, preparing for higher education and preparing for a career can go hand in hand:

> *Every high school graduate should find viable ways of pursuing both a career and a postsecondary degree or credential. For too many of our youth, we have treated preparing for college versus preparing for a career as mutually exclusive options.*[viii]

Some individuals and groups have criticized the vision put forth in *Pathways to Prosperity*, saying that its recommendations are calling for different standards, "tracking" students, and denying opportunities for some students, particularly those who are poor and of color. But Gary Hoachlander, a specialist in career and technical education, has argued just the opposite. Hoachlander points out increased options in high school promote student choice, not student tracking:

> *We must recognize that there are many different ways for high school students to pursue and achieve excellence. Imposing a uniform academic experience on everybody, simply to avoid the specter of tracking, is not in the best interest of all students. What engages and motivates some students will not excite and move others. Until we accept this simple fact, we will not make much progress on our nation's shameful dropout rate and substandard achievement.*

> *Let's be clear: This is not about accepting low standards for some and high standards for others. Rather, it is about recognizing that in our full and complex world, excellence and success take many forms—a single pathway is very much at odds with promoting widespread accomplishment for all students. Uniformity simplifies policymaking; it does not nurture deeper learning. It's time to move beyond the simplistic college-for-all-vs.-tracking debate and get about the very difficult business of designing and delivering high schools that engage and motivate young people.*[ix]

As Hoachlander points out, our nation's colleges already recognize that different students have different interests, aptitudes, and talents. Colleges respond to this diversity by offering students a variety of options in their pursuit of both general education and major areas of specialization. For Hoachlander, our challenge as a nation is to help high schools offer a similar variety of career pathways; he offers four principles for meeting the challenge of creating new high school pathways:

- *A pathway, by design, should prepare students for both college and career. The days are gone when someone could succeed with just a high school diploma. Everyone will need further education and career preparation.*

- *A pathway should prepare students for the full range of postsecondary options. "College" no longer just means a four-year postsecondary opportunity. It also includes community college, apprenticeship, and formal employment training.*

- *A pathway should connect challenging academics to the real world, helping students to better understand what they need to know and why they need to know it. Students deserve thoughtful and truthful answers when they ask, "Why do I need to know this?"*

- *A pathway must produce significant growth in student achievement—in academics to be sure, but also in communications, critical thinking, problem-solving, technological literacy, and other cross-disciplinary areas needed for success in the modern world.[x]*

Not only are these principles good for students; they are also doable: Hoachlander's state of California has already adopted them to guide its own high school reform efforts.

So, How Can High Schools Help Students Who Want to Be STEM Technicians?

This book outlines three strategies to improve high school preparation for future STEM technicians:

1. Middle and high school guidance counselors must be equipped to identify students who have the unique abilities and interests to become STEM technicians. Counselors will need to make these students and their parents aware of their unique abilities and provide students with the information and incentives that will motivate them to pursue this pathway.

2. High schools must provide alternative curriculum choices. STEM programs must provide an alternative pathway, to begin in the student's junior year, that articulates to an AAS STEM program at a community or technical college.

3. High schools and community and technical colleges must work together to create seamless 4+2 (secondary/postsecondary) Career Pathways that will allow, encourage, and prepare students to pursue studies that will lead to AAS degrees. Deficiencies in mathematics are the greatest cause of student attrition in the first semester of college STEM technician programs. These pathways, therefore, must ensure that students are proficient in mathematics up to the pre-calculus level by the time they graduate from high school.

Are we "tracking students" with this recommended approach? Absolutely not, if the pathway for STEM technicians only splits after the sophomore year in high school. It is at this time when most applied learners begin to lose interest or become discouraged and drop out of the STEM program.[xi] Offering an alternate pathway actually provides the opportunity to retain more students in STEM programs. And it also allows students to take additional dual credit courses in their technical field, which will provide them one or two semesters' worth of postsecondary credits. This is a huge cost savings that will enable more students with limited financial resources to "get their degree."

In this brief introduction, I have attempted to highlight a serious problem in technical education that is affecting our nation's workforce and its ability to innovate and compete. The solution to this problem lies in part with two existing initiatives: the NSF/ATE program and high school STEM programs. Chapters 2 and 3 describe these two initiatives in detail. Chapter 4 offers a strategy that high schools and technical colleges could use to prepare highly qualified future technicians. Chapter 5 describes dual-credit courses, a key element in the solution.

Most of the strategies and curriculum models to be used in creating career pathways for STEM technicians have already been at least partially developed and tested, and are currently in place in several NSF/ATE projects and Centers. Chapters 6–13 provide examples of these existing STEM Pathways in critical technologies that colleges are addressing through NSF/ATE initiatives.

If we are to solve the problem of "not enough STEM technicians," we must recognize the need for STEM program alternatives, create a collective vision for establishing them, and enact local and state policies that will support these changes. These proposed efforts are described in chapters 14 and 15.

And once again, we will need the characteristics, cited at the beginning of this chapter, that made our country great and its people safe and prosperous:

Vision—Courage—Hard Work
A Public Education System that is Second to None
Technological Innovation—Creativity

[i] Committee on Prospering in the Global Economy of the 21st Century: An Agenda for American Science and Technology, National Academy of Sciences, National Academy of Engineering, Institute of Medicine, *Rising Above the Gathering Storm: Energizing and Employing America for a Brighter Economic Future* (The National Academies Press, 2007).

[ii] Members of the 2005 "Rising Above the Gathering Storm" Committee; Prepared for the Presidents of the National Academy of Sciences, National Academy of Engineering, and Institute of Medicine, *Rising Above the Gathering Storm, Revisited: Rapidly Approaching Category 5* (The National Academies Press, 2010).

[iii] See chapter 3 for more information about STEM academies.

[iv] See Dale Parnell, *Why Do I Have to Learn This? Teaching the Way People Learn Best* (Waco, TX: Center Occupational Research and Development, 1995).

[v] The availability and quality of many of these programs are the result of National Science Foundation Advanced Technological Education (NSF/ATE) grants that support curriculum development, laboratory equipment, and professional development for college faculty. Chapter 2 offers a detailed discussion of the NSF/ATE program and the colleges it supports.

[vi] Harvard Graduate School of Education, *Pathways to Prosperity: Meeting the Challenge of Preparing Young Americans for the 21st Century*, February 2011.

[vii] Dan Hull, *Career Pathways: Education with a Purpose* (Waco, TX: CORD, 2005).

[viii] *Pathways*, 24.

[ix] Gary Hoachlander, "Toward a New Vision for American High Schools," *Education Week*, February 18, 2011.

[x] Hoachlander, "Toward a New Vision."

[xi] Chapter 4 discusses students' experiences and high school STEM curricula at greater length.

2 Chapter

Curriculum and Teaching Strategies for STEM Technicians
The NSF Advanced Technological Education Program

Elizabeth J. Teles[i]
Teles Consulting, LLC

In 1992, the U.S. Congress passed the *Scientific and Advanced-Technology Act (SATA)* to encourage, guide, and support our nation's community and technical colleges in preparing science and engineering technicians to support U.S. employers in advanced and emerging technical fields. This act has resulted in a consistently increasing federal fund for Advanced Technological Education (ATE), administered by the National Science Foundation (NSF). ATE grants support the design and improvement of our country's community and technical college infrastructure for delivering STEM technician education at the associate degree level.

The Emergence and Growth of ATE

SATA's purpose was

> to establish a national advanced technician training program, utilizing the resources of the Nation's two-year associate-degree-granting colleges to expand the pool of skilled technicians in strategic advanced-technology fields, to increase the productivity of the Nation's industries, and to improve the competitiveness of the United States in international trade, and for other purposes.[ii]

The act's authors listed several reasons that passage of the act was necessary; these reasons are as vital and important today as they were then:

1. the position of the United States in the world economy faces great challenges from highly trained foreign competition;
2. the workforce of the United States must be better prepared for the technologically advanced, competitive, global economy;
3. the improvement of our work force's productivity and our international economic position depend upon the strengthening of our educational efforts in science, mathematics, and technology, especially at the associate-degree level;
4. shortages of scientifically and technically trained workers in a wide variety of fields will best be addressed by collaboration among the Nation's associate-degree-granting colleges and private industry to produce skilled, advanced technicians; and
5. the National Science Foundation's traditional role in developing model curricula, disseminating instructional materials, enhancing faculty development, and stimulating partnerships between educational institutions and industry, makes an enlarged role for the Foundation in scientific and technical education and training particularly appropriate.

Over the years, the Advanced Technological Education (ATE) program has stayed true to its mission and has worked closely with community and technical colleges and their partners to help prepare the competitive United States workforce envisioned in this legislation.

Originally, the NSF was concerned that SATA's efforts would be too narrowly focused on training for specific jobs that might not be available over the long term. The NSF did want to play a role in fostering technical education, however. In testifying before Congress, Dr. Luther Williams, the then NSF Assistant Director for Education and Human Resources, said that the NSF's resources were "best employed in efforts to improve broad and general technical skills, backgrounds, and competencies, rather than in supporting improvements in training students for specific jobs which may have a short life span."[iii]

When the legislation passed in 1992, the challenge was to create a program that fulfilled the requirements of the SATA bill; built on the strengths and mission of NSF; and complemented, but did not duplicate, workforce programs at other agencies that provided funding to community and technical colleges. Programs at the Department of Education, such as Tech Prep, were already funding occupational programs that established partnerships between high schools and community and technical colleges. Programs at the Department of Labor, such as the Job Partnership Training Act, were providing funds for short-term training for disadvantaged and displaced workers.[iv] The ATE program needed to build on these existing efforts to fulfill Dr. Williams' vision of preparing highly qualified science and engineering technicians with broad technical skills and knowledge and a strong foundation in mathematics and science.

The initial planning for ATE began in a workshop called *Gaining the Competitive Edge: Critical Issues in Science and Engineering Technician Education.*[v] This workshop brought together selected educators and business and industry leaders to plan ATE. The educators came from the full range of educational institutions, including high schools, community and technical colleges, and four-year colleges and universities. The industry leaders included both managers and technicians. These leaders identified two types of advanced engineering and science technicians:

- Technicians who work in teams with scientists and/or engineers (for example, in research and development, manufacturing, and service organizations)
- Technicians who do not usually work directly with the scientists and engineers but provide support for their efforts (for example, field service technicians, IT professionals, and technicians who provide services for outside organizations).

The workshop clearly stated that technicians are not "junior professionals" whose work requires a less rigorous, more "applied" education (a view that often leads educators to develop curricula that are, at best, irrelevant and at worst, a barrier to entry). Rather, technicians are members of unique occupations with their own bodies of knowledge that are not a subset of those of engineering. Technicians work on or with complex technologies, need contextual knowledge and hands-on experience, and require formal knowledge in science and mathematics that forms the basis for the context of practice.

Since this workshop, a third type of technician has emerged. This technician combines technical knowledge and skills with other education and training, for example, in business, management, or teaching.

ATE Technicians Help Make U.S. Employers More Competitive

From its beginnings, the ATE program was designed to educate a vital arm of the technical workforce team needed to keep the United States competitive. It supported projects that were "guided by a coherent vision of technological education—a vision that recognizes the needs of the modern workplace, mathematics and science as the foundation for the required technology, students as lifelong learners, and the articulation of educational programs at different levels."[vi] The ATE program:

- Assures that associate degree programs help students acquire core skills and knowledge, basic technical skills that allow them to learn new skills as the jobs change, and advanced technical skills that provide hands-on experiences and knowledge so that students are employable and immediately useful to business and industry.
- Works in partnerships with business and industry to assure that ATE-supported programs produce well-prepared workers with adaptable skills.
- Creates articulated educational pathways for potential technicians from high schools to technical programs in community and technical colleges.
- Provides opportunities for graduates of two-year college technician programs to continue their education at four-year institutions to gain additional technical and management skills. These opportunities allow technicians to further advance in their fields or to become technical managers.

ATE technicians are different from craft workers. They are knowledge workers in high-technology fields. ATE projects and centers work closely with employers to assure that ATE technicians have both the skills and knowledge they need to be immediately valuable to employers and the ability to adapt and develop as jobs change. ATE is not about short-term training; it is about preparing students for careers with opportunities for advancement. The ATE supports programs that represent joint efforts between community and technical colleges and business and industry. Research has shown not only that these joint programs work, but also that they work better than technical programs that colleges create on their own and then must sell to employers. These programs are win-win situations for both colleges and employers. The following examples illustrate:[vii]

From an employer:	About a technician:
A telecommunications employer said he found the ATE program valuable because it reduced the amount of additional training he needed to provide employees. He said, "I need an employee who I can send for additional training one week every six to twelve months, not one that needs six months of additional training every six months."	A chemical technician began to work for a chemical company upon graduating from an ATE program that was jointly developed between the company and the college. She was immediately a valued employee and ready for the propriety training she needed. The company then paid for a four-year college education that allowed her to earn advanced skills and a promotion. Moreover, they supported her in a master's degree program that allowed her become a technical manager.

Figure 2.1. Employers and students benefit from ATE programs

The Growth of ATE and Its Influence on College Programs

In 1992 with SATA, Congress authorized up to $35 million annually for creation of the ATE program to support ten Centers of Excellence and additional projects. In 1994, Congress appropriated $14.5 million in ATE initial funding, adding over $11 million to the NSF request of $3.5 million. Over the last eighteen years, the ATE annual budget has grown from $14.5 million to $64 million. Congress supports the ATE program because members see the value of partnerships among two-year colleges, high schools, four-year colleges and universities, and business and industry. These partnerships provide well-qualified workers for the high-performance technology workplace and help build local and regional economies. While ATE is not a large workforce-development program by Departments of Education and Labor standards, it is a large program for NSF: it operated with a $64 million budget in the 2011 fiscal year. And ATE has a unique niche among workforce programs in that it focuses on preparing highly qualified science and engineering technicians, as opposed to providing short-term training or preparing craft workers or paraprofessionals in other career fields. In its eighteen years of existence, the ATE program has provided over $720 million to community and technical colleges and their partners and has provided additional amounts in excess of $100 million through co-funding with other NSF programs. In addition, ATE projects and centers are leveraging support from other sources, such as business and industry, state governments, and foundations. A conservative estimate, based on annual surveys, is that these groups have provided over $500 million in additional funding.

Since 1994, ATE has supported thirty-nine Centers of Excellence and almost one thousand projects. Grants went to more than five hundred community and technical colleges in every state plus Puerto Rico. In 2010, more than 70,000 students were involved in ATE courses and programs. Of these students, 60 percent were enrolled in community or technical colleges, 33percent were enrolled in high schools (often in dual enrollment classes that prepared them for community or technical colleges), and 7percent were either enrolled in four-year institutions or working in business and industry and returning to a community or technical college to upgrade their skills and knowledge. In addition, ATE provided professional development to over fifty-eight thousand educators to help them provide better education to current and future technicians.[viii]

The Scope of the ATE in Addressing Employers' Broad Technical Needs

The ATE program supports a broad range of technician programs closely aligned with the needs of local, regional, and national businesses and industries. In 2010, ATE supported centers in these areas:

- Advanced manufacturing technologies
- Agricultural, energy, and environmental technologies
- Biotechnology and chemical processes
- Electronics, microelectronics, and nanotechnology
- Engineering technologies in fields such as marine technology, optics and photonics, transportation, and space
- Information, geospatial, and security technologies
- Learning, evaluation, and research[ix]

In addition to supporting centers that focus on specific technologies, the ATE also supported interdisciplinary projects that provided professional development for faculty and teachers and promoted student recruitment and retention.[x]

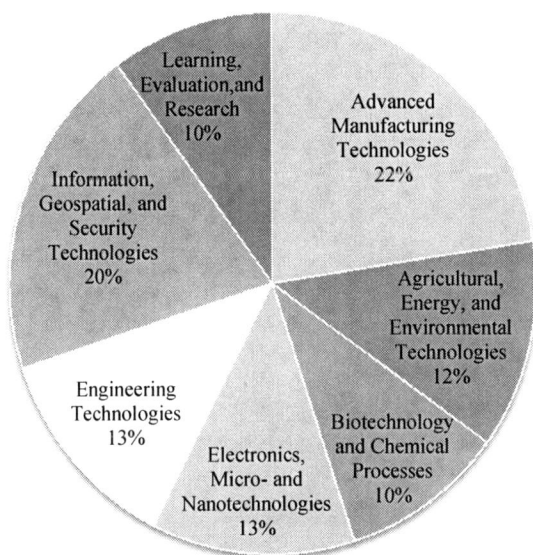

Figure 2.2. ATE Centers

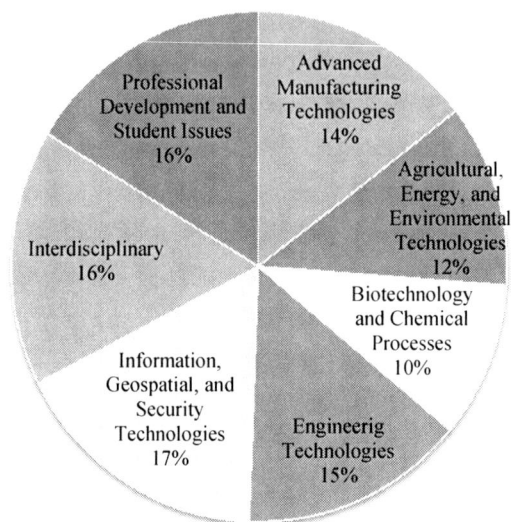

Figure 2.3. ATE Projects and Centers

Community and technical colleges and their partners collaborate in many ways to prepare highly qualified and flexible science and engineering technicians:[xi]

- The Automotive Manufacturing Technical Education Consortium (AMTEC) is a vital part of the Kentucky Community and Technical College System. AMTEC was originally funded as a project in partnership with Toyota and now brings together thirty community and technical colleges with thirty-four employers in twelve states. AMTEC is committed to producing a new kind of technician, a multidisciplinary problem solver who thinks in systems and is an integral part of continuous improvement. Working closely with the K–12 system, AMTEC provides a seamless pathway to a technical community college associate degree and into career-long learning in the workplace. Current workers take advantage of the modularized curriculum to upgrade skills and earn college credits, certifications, and degrees.
- Bio-Link, the Next Generation National ATE Center for Biotechnology and Life Sciences, prepares students for entry-

level positions in both large biotechnology companies and smaller biotechnology start-ups.

- CyberWatch works to increase the quantity and quality of information assurance (IA) workers. The consortium has grown to include fifty-four colleges and universities and more than thirty business and government employers. Close connections to the National Security Agency (NSA), the Department of Homeland Security, industry associations, and private cybersecurity providers assures that programs and curricula meet national standards. Six community colleges in particular have been certified by the NSA as National Centers of Excellence in Information Assurance Two-Year Education.

ATE's Broad Influence, Leadership, and Leveraged Support

The ATE program has provided leadership by anticipating changes in the skills and knowledge needed in the U.S. technical workforce in diverse fields:[xii]

- **Information Technology.** The Northwest Workforce Center for Emerging Technologies (NWCET) produced skill standards that community and technical colleges all around the country used as the basis for their IT programs. Microsoft found these programs so valuable that in 2003, it helped establish the Working Connections project, through which it provides additional support for community and technical colleges to develop and implement IT programs.

- **Cybersecurity.** In 2002, NSF sponsored an American Association of Community Colleges (AACC) workshop called *Protecting Information: The Role of Community Colleges in Cybersecurity Education.* The workshop developed recommendations on the roles that highly skilled technicians can play in this emerging field. ATE later supported cybersecurity education through three Centers of Excellence; together, these centers set standards for associate degrees in cybersecurity.

- **Biotechnology.** For over a decade, Bio-Link, an ATE Center of Excellence in biotechnology, has worked with other ATE biotechnology centers and projects to set expectations and train faculty for biotechnology programs in community and technical colleges. Bio-Link and other centers continually update their programs, incorporating new knowledge as the industry changes. In 2008, NSF and the American Association of Community Colleges sponsored a workshop on biotechnology education called *Educating Biotechnicians for Future Industry Needs.*[xiii]

An Example of Employer Need in an Emerging Technology

"Our companies need workers with knowledge of molecular biology, biochemistry, and cell culture. They need workers who can perform basic research, operate standard lab equipment, understand instrumentation, and follow lab protocols."

–James Greenwood, chief executive officer of the Biotechnology Industry Organization

---

ATE Growth in Supporting Alternative Energy

As the ATE program grows, it continues to look to the future to help shape community and technical college programs. ATE has always supported projects focusing on energy, but in the past, these projects were often add-ons to existing electronics programs. But the ATE program has assumed a leadership role in energy education as new jobs have emerged for advanced technicians in alternative energy fields (e.g., wind, solar, and geothermal) and in sustainable construction and maintenance in more traditional fields (e.g., electrical, natural gas, nuclear, and coal). In 2009 and 2010, NSF collaborated with other federal and state agencies to hold a series of seven regional energy conversations, culminating in a National Summit in December 2010.[xiv] Participants identified jobs that energy technicians currently perform, including energy assessment; energy-efficient building construction; project engineering and implementation; exploration, generation and utility-scale construction; operations and maintenance; regulatory affairs; transmission and distribution; transportation services; and the purchase and sales of energy.[xv] Industry representatives called for technicians with some work experience and a good, solid, fundamental understanding of science and related safety issues. They wanted technicians with skills in mathematics, science, data analysis, mechanics, and information technology.

In the 2010 and 2011 fiscal years, nearly 25 percent of the ATE portfolio (twenty projects and centers) supported the development of model programs in energy-related fields. These awards enhanced existing programs for natural gas and nuclear technicians, developed new programs for technicians in alternative energy fields, and created programs for technicians in sustainable energy technologies.[xvi] Three examples demonstrate the diversity of these efforts:

- In Wyoming, Laramie Community College is improving wind energy programs for technicians by increasing academic rigor and relevance. Core competencies include electricity, hydraulics, fluid power, and mechanical drive systems; these are also core competencies in other energy programs. The program combines wind-energy-specific knowledge and skills with core academic competencies and industry-driven, performance-based standards for integrated systems.

- In Oregon, Linn-Benton Community College is expanding its existing mechantronics associate degree program to include a Green Energy Certificate that blends energy generation, energy conservation, and practical physics.

- In Massachusetts, Holyoke Community College is collaborating with numerous regional partners to create associate degree programs and certificates that prepare students to enter the clean energy job market as auditors, builders, and installers for wind, solar, and geothermal companies.

Challenges in Reaching Adequate Numbers of Technician Graduates

ATE programs at community and technical colleges have great success in finding student internships and placing graduates in excellent jobs—in fact, most of them have a greater-than-90 percent placement rate. But many of the programs suffer from low enrollment and high student attrition, depend heavily on more mature students who wish to change jobs, or serve incumbent workers. As a result, the college infrastructure developed and sustained by ATE is not fully serving employers' technician needs. To address this problem, virtually all colleges working with ATE projects and centers are exploring innovative ways to partner with local feeder high schools through dual and concurrent enrollment programs, career exploration, and technical experiences for high school students. But if the United States is to be more competitive, ATE programs must enroll and graduate many more students. In a survey conducted by Western Michigan University, 56 percent of ATE principal investigators identified difficulty in recruiting students as their number one challenge.[xvii] All of this points to the need to create a more fully developed and effective high school pipeline to science and engineering technician programs.

In background information, ATE principal investigators cited the following reasons for inadequate enrollments levels in community and technical colleges:

- Many students lack important mathematics skills, as demonstrated by low placement test scores and high failure rates in mathematics courses.

- Many students either dropped out of high school or performed poorly in high school; these students typically took an inadequate number of high school mathematics and science courses or took these courses long before entering the community or technical college.

- Students who come to community and technical colleges without specific educational or career goals are placed in general education programs when they should instead be given guidance in selecting a major.

- Colleges enroll huge numbers of students in developmental mathematics courses without regard for students' career goals.

- Colleges do not provide students the opportunity to take technology courses for credit in their majors while concurrently taking developmental courses.

- High school counselors and teachers often do not provide appropriate course-selection guidance; as a result, many high school students do not take the mathematics and science courses necessary to prepare them for technician programs.

- High schools rarely provide students with technological education or technical experiences to prepare them for college-level technical programs.
- Many teachers and counselors do not know very much about careers in technical areas and discourage students from entering technical programs.
- Students and their parents often do not know very much about technical jobs and aren't aware of the excellent opportunities for rewarding technical careers.
- Counselors, teachers, parents, and students often misperceive technician jobs as poorly paid, uninteresting, and dirty, and believe that that graduates are just cheap labor for industry.
- Many students lack study skills and necessary levels of maturity.

While these problems seem huge, community and technical colleges and feeder high schools have developed many models that work. Through its *Achieving the Dream* program, the Lumina Foundation has been helping community and technical colleges move students from developmental education courses into credit programs.[xviii] The Carnegie Foundation for the Advancement of Teaching is working with nineteen community colleges in five states to implement *Statway,* an alternative developmental mathematics curriculum that leads students into credit-level courses and programs.[xix] The philosophy of *Statway* is (a) that statistics is not the same mathematics course that students may have failed before and (b) that the principles of statistics are relevant for many careers.

Promising Practices to Improve the High School Pipeline

While the ATE program has not solved the problems associated with low enrollments in technician programs, it has developed many promising practices.[xx] Most importantly, ATE projects and centers have built alliances with high schools to help prepare students for college and careers. College faculty and high school teachers work together to help solve many of the problems raised above and have achieved huge successes. Different schools face different challenges, and the strategies for addressing them necessarily vary. However, many challenges relate to the three themes discussed below. Under each challenge, I describe four different ways that schools have responded to that challenge. These examples are not exhaustive; instead, they show the variety of possible solutions that schools can adopt.

Challenge 1: Students come to the community or technical college unprepared or underprepared.

    Solution A: Colleges place cohorts of students into accelerated "academies" within the college. Students enroll full time for one or two semesters in programs designed to prepare them for college work. Programs not only teach academic coursework (including mathematics), but also build self-efficacy and improve students' professional and study skills. For example, the vision of the Academy for College Excellence (formerly the Digital Bridge Academy) at Cabrillo College and other partner colleges is to "increase the number of students who emerge from community college prepared for a knowledge-work professional career with a future."[xxi]

    Solution B: Students enroll in bridge programs at the community or technical college for specific technical programs. For example, the Bridge to Biotechnology at the Community College of San Francisco and adapted by many colleges around the country is designed to integrate biotechnology laboratory skills into developmental mathematics, science, and English courses.[xxii]

    Solution C: Students co-enroll in developmental courses while taking credit courses. If enough students from a particular program enroll, the college designates special sections of developmental mathematics for its majors; these special sections include examples or experiences targeted to those students. For example, in some teacher-preparation programs for elementary teachers, students enrolled in developmental mathematics courses tutor young children in mathematics.

    Solution D: Students enroll in dual- or concurrent-enrollment mathematics, science, or technology courses while in high school. These courses are taught by college faculty or high school teachers who are certified to teach college courses; they expose students to college-level learning and provide seamless transfer.

Challenge 2: Students have insufficient opportunities to take technical courses and participate in technical experiences while in high schools.

Solution A: Students take dual-enrollment technology-related courses at the community or technical college while they are still in high school. Students generally receive college credit for these courses after they have completed one or more additional technology courses at the college.

Solution B: Students enroll in college technology classes while they are still in high school. They earn college credit by taking courses at the community college or at technology centers after school, in the summer, or on weekends.

Solution C: High school career academies prepare students for college-level work and expose them to a variety of careers, including many in high-technology fields. Career academies usually involve a small learning community within a high school; a college-prep curriculum with a career theme; and partnerships with employers, the community, and higher-education institutions.[xxiii]

Solution D: Community and technical colleges, high schools, and other partners work together to provide a wide variety of technical experiences for students. These include summer camps and after-school activities as well as preparation for and participation in competitions such as First Robotics and CyberPatriot.

Challenge 3: Students receive poor or insufficient career guidance.

Solution A: High school programs include career exploration and guidance; often, these programs begin in middle schools. These programs may take the form of required courses, individual education plans, and integration of career guidance into existing courses.

Solution B: Specialized STEM counselors in high schools and community and technical colleges guide students to careers in STEM fields. They encourage students to take appropriate courses and participate in activities that assure that students will be able to enter STEM college programs.

Solution C: Industry mentors and coaches work with students individually and in groups to assure that high school students understand the world of work and what the good jobs of today require.

Solution D: Students in high schools and colleges build online communities based on shared interests; colleges use online resources to expose students to emerging technologies and technical careers.

In 2010, twenty-four thousand high school students were enrolled in ATE-supported courses and activities.[xxiv] But ATE cannot solve the problem of low enrollments in technical programs alone. At best, it can help initiate efforts and support community and technical colleges and high schools as they collaborate to create models that work. The ATE program is about advanced technician education, but for technician programs to have maximum impact, students must come to them ready to succeed.

So what do college technician programs need from their supporting (feeder) high schools? ATE and other college technician programs need high schools, in collaboration with community and technical colleges, to:

- Increase visibility: help students learn about career opportunities in emerging technologies.
- Identify students who may be interested in careers as technicians, encourage them to consider these careers, and prepare them for college-level technician programs.
- Develop and offer courses that build on the unique abilities of students who enjoy working with technology.
- Build on students' interest in computers, televisions, and hand-held devices such as iPhones and Blackberries.
- Offer mathematics and science courses that are relevant to and appropriate for a broad range of technical programs.
- Incorporate technical experiences into classes and offer after-school and Saturday programs that feature technical experience.
- Work with employers to develop summer and after-school work opportunities in hands-on environments.

Summary

The federally funded NSF/ATE program continues to design, encourage, and support a broad range of community and technical colleges' associate degree technician programs. These programs meet the demands of employers involved with new and emerging technologies that serve our country's defense, energy, and economic needs. However, the infrastructure that ATE has put in place is not being used to capacity—and employer's needs for technicians are not being adequately met—because colleges are not registering and retaining a sufficient number of well prepared students.

The best solution to the problem of low enrollment is a strong pipeline that will lead students from high school directly to college-level technical programs. ATE has devised and tested promising practices to build this high school pipeline by helping colleges partner with local high schools. But these partnerships are severely limited because most high schools focus their STEM curricula on preparing students for engineering and science programs at four-year colleges and universities. For example, the topics in the standard high school mathematics curriculum (Algebra I, Geometry, Algebra II, Precalculus) have been chosen to prepare all students to take calculus in either high school or college. We must create more and larger partnerships that will allow potential technician students to articulate into associate-degree STEM programs at community and technical colleges. These partnerships must develop more effective strategies to identify and encourage potential technician students, and they must create a curriculum to prepare these students for a smooth transition into college-level technical programs.

---

[i] Elizabeth J. Teles was formerly the ATE co-Lead at the National Science Foundation; the statements and opinions in this chapter are those of the author and do not necessarily reflect those of the NSF.

[ii] U.S. Senate, *Scientific and Advanced-Technology Act of 1992*, 102nd Cong., 2d sess., 1992, S. 1146.

[iii] *Remarks of Dr. Luther S. Williams*, Assistant Director for Education and Human Resources, National Science Foundation, at September 17, 1991 Hearing of the Subcommittee on Technology and Competitiveness, Committee on Science, Space, and Technology, United States House of Representatives.

[iv] Robert Guttmann, "Job Training Partnership Act: New Help for the Unemployed," *Monthly Labor Review* 106, no. 3 (March 1983).

[v] National Science Foundation, *Gaining the Competitive Edge: Critical Issues in Science and Engineering Technician Education*, NSF 94-32 (April 19, 1994).

[vi] National Science Foundation, *Advanced Technological Education (ATE) Program Solicitation*, NSF 00-62 (2000).

[vii] "From an employer" source: personal communication to Elizabeth Teles at Springfield Technical College. "About a technician" source: personal communication to Elizabeth Teles at *Faces of Success*, ATE Principal Investigators Conference 2008.

[viii] Lori Wingate, Carl Westine, and Arlen Gullickson, *Advanced Technological Education Survey 2011 Fact Sheet*, July 2011, http://evalu-ate.net/downloads/2011%20Fact%20Sheet.pdf

[ix] Madeline Patton, ed., *ATE Centers Impact 2008–2010: Partners with Industry for a New American Workforce* (Tempe, AZ: Maricopa Community Colleges, 2008–2010).

[x] See note 8 above.

[xi] Examples are from Madeline Patton, ed., *ATE Centers Impact 2007–2011* (Tempe, AZ: Maricopa Community Colleges, 2007–2011) and Chris L. S. Coryn, Arlen R. Gullickson, and Liesel A. Ritchie, *Advanced Technological Education Program Evaluation Project Implementation: Challenges and Resolutions Briefing Paper 5* (Evaluation Center at Western Michigan University, 2006).

[xii] See note 11 above.

[xiii] Madeline Patton, *Educating Biotechnicians for Future Industry Needs* (Washington, DC: Community College Press, 2008); the quote from James Greenwood is on page 8.

[xiv] Madeline Patton, "National summit examines energy technician education," *Community College Times*, December 15, 2010.

[xv] Advanced Technology Environmental and Energy Center, *Defining Energy Technologies and Services (Bettendorf, IA,* 2008).

[xvi] Advanced Technological Education 2011, *What Has Been Funded (Recent Awards Made Through This Program, with Abstracts),*

http://www.nsf.gov/funding/pgm_summ.jsp?pims_id=5464&org=DUE&fro
m=home; National Science Foundation, *Advanced Technological Education
(ATE) Program Solicitation*, NSF 11-692 (2011).
[xvii] See Coryn, Gullickson, and Richie, note 11 above.
[xviii] *Achieving the Dream: Success is What Counts*,
http://www.achievingthedream.org/docs/SUCCESS-counts-FINAL-11.6.pdf
[xix] "Statway Overview," http://www.carnegiefoundation.org/statway
[xx] Madeline Patton, "ATE Develops Student Recruitment and Retention
Strategies," *TECHcitement*, American Association of Community Colleges,
2008; Madeline Patton, ed., *ATE Projects Impact: Partners with Industry for
a New American Workforce* (Community College Press, 2008); Laura
Brooks, Keith MacAllum, and Amanda McMahon, *The Bridge to
Employment Initiative: A Decade of Promising School-to-Career
Partnerships* (Academy for Academic Development, 2005).
[xxi] Academy for College Excellence, "Our Mission,"
http://academyforcollegeexcellence.org/mission-vision/
[xxii] Madeline Patton, ed., *ATE Centers Impact 2008–2010* (Tempe, AZ:
Maricopa Community Colleges, 2007–2011).
[xxiii] Florida Department of Education, "About Career Academies,"
http://www.fldoe.org/workforce/careeracademies/ca_home.asp
[xxiv] See note 16 above.

High School STEM Initiatives

L. Allen Phelps
University of Wisconsin-Madison

Introduction

As the United States develops new education policies for preparing the next generation of a high-skill, high-wage, and high-demand workforce, high school STEM initiatives will play a vital role in addressing major societal challenges worldwide. Making solar energy economical, restoring urban infrastructure, and securing cyberspace are three of the fourteen grand challenges cited recently by the National Academy of Engineering.[i] Finding solutions to these and other challenges at the local, national, and global levels will require new skills and talents that most high school leaders and educators are committed to teaching.

Many communities are in the early stages of developing a foundation for postsecondary STEM education. Of the nation's thirty thousand public schools with secondary grades (grades 7-12):

- Approximately 25 percent or (7,500 schools) operate career academies, and about 40 percent of these academies (about 3,000) have a broadly defined STEM theme.[ii]

- In 2011, approximately 12.5 percent of the nation's 2,004 high school graduates had completed a two-credit concentration in STEM-career-cluster courses (e.g., engineering technology, health sciences, or information technology).[iii]

The central premise of this book is the need for a widely and deeply integrated STEM education to prepare tomorrow' technicians, engineers, and scientists. In this chapter, I describe some promising practices as well as the major challenges that high schools must confront in preparing students to succeed in community and technical college STEM pathways.

Since at least the early 1950s, America's high schools have been at a critical crossroads of a continuous debate on the nation's economic future. In each era of economic innovation—from the agricultural to the manufacturing through the knowledge economies—high schools have been challenged to serve a wider segment of the nation's youth with increased general education requirements as well as new academic, career, and technical specializations.

During most of the twentieth and early twenty-first centuries, American communities have centered their conversations about education on two driving questions: What should the common school curriculum include? And what part of the curriculum should be focused on enabling students to succeed in life after adolescence? Following World War II, education advocates argued for James Conant's comprehensive high school curriculum, which included four units of English for all students, as well as three years of math, science, and foreign language for the small minority of those in the "college prep" track.[iv] The vocational-technical track was available for students who planned to work in the thriving manufacturing, agricultural, and business sectors, while the general-education track allowed students to explore a variety of other interests.

By the mid-1980s, some educational scholars were characterizing high schools as shopping malls that offered a wide range of specialized courses to a diverse population that included students with learning and physical disabilities and students whose native language was not English.[v] By 1987, a lack of academic rigor and coherence in the high school curriculum had placed the "Nation at Risk", according to the National Commission on Excellence in Education.[vi]

To address the growing demands of the information- and knowledge-driven economy, the commission argued, all students should complete the "new basics":

- four years of English;
- three years of mathematics, science, and social studies;
- one-half year of computer science.

A Nation at Risk also recommended changes in the academic content and instructional strategies for these courses. Mathematics courses, for example, should enable students to apply mathematics and scientific understanding in everyday situations and understand the social and environmental implications of scientific and technological development. In addition, the commission deemed it essential to develop national and state academic curriculum standards and assessments that would upgrade the capacity of all high schools to prepare students for the fast approaching twenty-first century. During the 1990s and 2000s, states and professional associations focused much of their high school improvement efforts on developing new curriculum and teaching standards, along with new state assessments.

Over the past decade, postsecondary education has become increasingly important due to a variety of new economic forces and workplace innovations, including global competition, marketplace research and customization, and multilingual marketing. In 2007, the authors of the National Academy's seminal report, *Rising Above the Gathering Storm*, noted:

> The prosperity the United States enjoys today is due in no small part to investments the nation has made in research and development at universities, corporations, and national laboratories over the last 50 years. Recently, however, corporate, government, and national scientific and technical leaders have expressed concern that pressures on the science and technology enterprise could seriously erode this past success and jeopardize future U.S. prosperity. Reflecting this trend is the movement overseas not only of manufacturing jobs but also of jobs in administration, finance, engineering, and research.

In addition to calling for incentives to expand and strengthen the math and science teaching workforce nationally, *Rising Above the Gathering Storm* made three recommendations that highlighted the substantial changes needed to improve high school STEM teaching and learning:

1. Enlarge the pipeline of students who are prepared to enter college and graduate with a degree in science, engineering, or mathematics by increasing the number of students who pass Advanced Placement and Internationl Baccalaureate science and mathematics courses.
2. Create *statewide specialty high schools* to foster the development of leaders in science, technology, and mathematics.

3. Implement *inquiry-based learning* through summer internships and research opportunities to provide valuable laboratory experiences for both middle school and high school students. [vii]

Demand for Middle-Skill Jobs

The increasing numbers of STEM-related careers call for a wide variety of skills and knowledge. While many STEM careers require bachelor's or graduate degrees, a large sector of the STEM labor market requires expertise in technical work and other specialties that need less than a bachelor's degree for workforce entry. In a detailed analysis of the U.S. Bureau of Labor Statistics (BLS) 2014 labor market projections, Holtzer and Lerman argued that "the demand for workers to fill jobs in the middle of the labor market—those that require more than high school, but less than a four-year degree—will likely remain quite robust relative to its supply, especially in key sectors of the economy."[viii]

This national analysis offers a number of key indicators that clearly suggest that the demand for STEM-related technical workers is stable and, in some occupations, is growing substantially:

1. Overall, BLS estimates that nearly half (about 45 percent) of all job openings in the next ten years will require middle levels of skill—significantly more than those that will require a high levels of skill or a baccalaureate (33 percent) or low levels of skill (22 percent).
2. Jobs in all the middle-skill categories combined (including positions requiring substantial on-the-job training) will generate about twenty-one million openings over the decade.
3. Jobs requiring "some college" will generate more job openings (15.5 million, or 28 percent of the total) than those requiring bachelor's degrees or higher (13.9 million, or 25 percent of the total).

Holtzer and Lerman offered the following projections about the demand for middle-skill-level jobs in STEM and STEM-related occupations:

1. The net growth in computer specialist jobs requiring less than a bachelor's degree will average about 20 percent and should generate more than one million job openings.
2. The net growth in a range of health care jobs with sub-baccalaureate education and training requirements will vary from 20 percent to 40 percent, with more than 1.5 million job openings.
3. Employment in five skilled construction trades is expected to grow by 10 percent to 15 percent and provide 4.6 million job openings, while jobs in installation/maintenance/repair and transportation will grow at similar rates and together generate more than four million additional openings.

Many of these observations and projections are reinforced in more recent reports by other prominent labor market economists. In a recent report entitled *The Undereducated American*, Dr. Anthony Carnevale argues that twenty million additional college graduates will be needed by 2025. He contends that when compared to other developed nations, "the United States has been underproducing college graduates for thirty years, leading to rising income inequality and other serious consequences."[ix] Carnevale's analysis includes the rising demand for associate-degree and workplace-certificate recipients.

Soon after taking office, President Obama announced a national college completion goal: by 2020, America will once again have the highest proportion of college graduates in the world. Reaching this goal will require significant public and personal investments in community colleges. Ensuring that 60 percent of the nation's twenty-four- to thirty-five-year-olds attain a college degree or credential by 2020 or 2025 (the Lumina Foundation's Goal) will require myriad new partnerships between local STEM employers and community and technical colleges.

Despite the lagging global and national economy, the demand for middle-skill workers will remain quite robust relative to its supply, especially in key sectors of the economy. A range of policies, including policies and programs to strengthen student interest and engagement in STEM careers, could help many students obtain education and training beyond high school for these middle-skill jobs in STEM and related fields. Quite clearly, increasing STEM engagement in high schools and community colleges would raise the earning levels and living standards of many families.

The Emergence of Secondary STEM Initiatives

As noted above, the rapidly changing economic context of the past decade has stimulated a number of new policies aimed at expanding the attention to STEM education in U.S. schools. The 2006 Amendments to the Carl D. Perkins Career and Technical Education Act (Perkins IV) required that local education agencies and postsecondary institutions receiving federal funds develop and offer Programs of Study (POS). Envisioned as *Career Pathways*, POS are secondary-to-postsecondary sequences of technical and academics courses that must:

- incorporate and align secondary and postsecondary education elements;
- include academic and career-and-technical-education (CTE) content in a coordinated, non-duplicative progression of courses;
- offer the opportunity, where appropriate, for secondary students to acquire postsecondary credits; and
- lead to an industry-recognized credential or certificate at the postsecondary level, or an associate or baccalaureate degree.x

Many states have moved forward with POS in the sixteen Career Clusters, which include seventy-nine pathways that have been formally recognized by national industry associations. According to the Association for Career and Technical Education (ACTE), six of the sixteen clusters are STEM-intensive. While ACTE has designated one official STEM cluster along with two associated pathways, STEM knowledge and skills are associated with several clusters. As ACTE notes,

> The elements of science, technology, engineering and math are integral parts of our nation's critical economic sectors, from health care to energy, and infrastructure to national security. STEM careers include not only those requiring a research-based advanced math or science degree, but a broad range of related occupations in areas as diverse as aquaculture,

CAREER PATHWAYS FOR STEM TECHNICIANS

automotive technology, accounting and architecture. More careers than ever before require a deep understanding of science, technology, engineering or math principles.[xi]

A detailed analysis crosswalk of the sixteen clusters suggests the following cross-cluster alignment of STEM-intensive clusters and pathways:[xii]

Agriculture, Food, and Natural Resource	Architecture and Construction	Arts, AV Technology, and Communications
• Environmental Service Systems • Natural Resources Systems • Plant systems • Power, Structural and Technical Systems	• Design and Pre-Construction • Construction	• Audio and Video Technology and Film • Telecommunications
Health Sciences	**Manufacturing**	**Science, Technology, Engineering, and Mathematics**
• Biotechnology Research and Development	• Health Safety and Environmental Assurance • Maintenance, Installation and Repair • Manufacturing Production Process Development • Production • Quality Assurance	• Engineering and Technology • Science and Mathematics

Figure 3.1. Cross-cluster alignment of STEM-intensive clusters and pathways

The sixteen States' Career Clusters were co-developed by educators and several national industry associations. The Career

Pathways are designed, where appropriate, to align with industry-recognized credentials. As the chart above indicates, STEM instruction serves multiple clusters and provides a critical gateway to filling technical, engineering, and scientific positions in the U.S. job market.

Several federal investments beyond the Department of Education have been key contributors to high school STEM innovation over the past decade. From 2002 to 2010, the National Science Foundation operated the Math-Science Partnership Program (MSP). Responding to growing national concern regarding U.S. children's mathematics and science skills, MSP awarded competitive grants to teams composed of institutions of higher education, local K–12 school systems, and their supporting partners. According to the NSF, funded Math-Science Partnerships have:

- Pioneered advancements in mathematics and science education.
- Provided innovation, inspiration, support, and resources to educators and students in local schools, colleges, and universities.
- Enabled partner organizations to cultivate and enhance their own strengths as they contribute to their MSP teams.
- Resulted in better-prepared students, and, ultimately, a better-prepared American workforce.

Over nearly a decade, these partnerships have brought together about 150 institutions of higher education, approximately 450 K–12 school districts, and a host of other stakeholders, including prominent business partners such as Pfizer, Ford Motor Company, Texas Instruments, and Xerox. The MSP operational framework (see below) offers a useful model for considering and leveraging the multiple STEM investments in a community or region.[xiii]

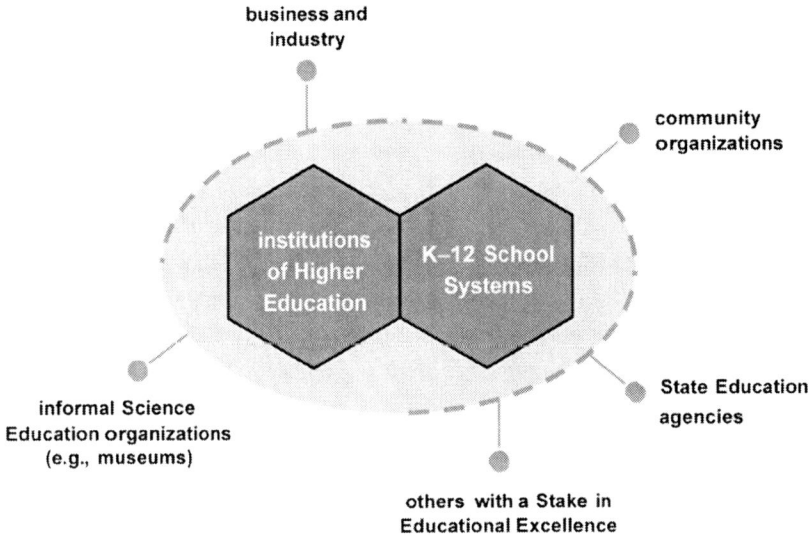

Figure 3.2. MSP Operational Framework

In 2001–02, the U.S. Department of Education launched a similarly named Math-Science Partnership Program under the No Child Left Behind (NCLB) Act. The focus and strategy for this MSP is substantially different from the NSF's MSP initiative. The Department of Education's MSP program focuses directly on teachers and aims to improve elementary and secondary students' achievements in mathematics and science by increasing instructional quality. All MSPs are aimed at improving instruction in high-need schools. To accomplish this goal, the program supports partnerships between high-need school districts and the mathematics, science, and engineering faculty of institutions of higher education.

Between fiscal years 2002 and 2008, Congress increased funding for MSPs from $12.5 million to $179 million. In 2008, 626 local partnerships were funded, which, in turn, served more than fifty-seven thousand educators nationwide. Each educator received an average of ninety-seven hours of professional development, which enhanced the quality of classroom instruction for over 2.8 million students.[xiv]

Without question, the Perkins IV and federal Math-Science Partnerships have fueled high school innovation over the past decade, as have other federal initiatives, such as the Department of Education's Small Learning Communities and Charter Schools programs.

Several regional organizations and foundation-supported networks have also advanced high school STEM education. The efforts of two major groups—the National Academy Foundation and Project Lead the Way—exemplify the considerable high school STEM innovation occurring nationally. Founded in 1982, the National Academy Foundation (NAF) provides leadership for preparing young people for college and career success. Over the past thirty years, NAF has created and refined a proven educational model which includes industry-focused curricula, work-based learning experiences, and business partner expertise for four themes: Finance, Hospitality & Tourism, Information Technology, and Engineering. Employees of more than 2,500 companies volunteer in classrooms, act as mentors, engage NAF students in paid internships, and serve on local Advisory Boards.[xv]

NAF's Academy of Engineering works with a variety of academic and industrial organizations to educate high school students in the principles of engineering. It provides content in the fields of electronics, biotechnology, aerospace, civil engineering, and architecture.[xvi] To date, NAF has helped establish fifty-four Academies of Engineering in U.S. high schools, in addition to 105 Academies of Information Technology.

The limited, non-experimental evidence on the impact of NAF's career academies is impressive:

- More than 90 percent of NAF students graduate from high school—compared with 50 percent in the urban areas where most NAF academies are located.
- Four out of five NAF students go on to college or other postsecondary education.
- 52 percent of NAF graduates earn bachelor's degrees in four years—compared with 32 percent nationally.
- Of those who go on to postsecondary education, more than 50 percent are the first in their families to go to college.
- 90 percent of students report that the academies helped them develop career plans. [xvii]

Established in 1997, Project Lead the Way (PLTW) was launched in several upstate New York high schools in response to a shortage of college students majoring in engineering. This national, nonprofit network now includes four thousand middle and high schools serving roughly 350,000 students annually. Through intensive two-week summer instructor training institutes hosted by affiliated engineering universities, PLTW has trained more than fifteen thousand certified technology education, math, and science teachers. According to the PLTW website, more than five hundred thousand students have enrolled in its courses.[xviii]

PLTW's pedagogical approach is embedded in case-, project-, or problem-based learning. Students undertake real-world projects to develop understanding and skills necessary to solve everyday-life problems, as well as the problems faced by engineering or health science firms. Schools can offer three PLTW programs, each of which is anchored in and aligned with national and state standards:

- PLTW Gateway To Technology (GTT) is a middle school program offered in six independent, nine-week units that helps students explore math, science, and technology. This activity-oriented program challenges and engages the natural curiosity of middle school students and is taught in conjunction with a rigorous academic curriculum.
- PLTW Pathway To Engineering (PTE) is a four-year high school sequence taught in conjunction with traditional math and science courses. PTE's eight courses, including Introduction to Engineering Design, Principles of Engineering, and Engineering Design and Development, provide students with in-depth, hands-on knowledge of engineering and technology-based careers. [xix]
- PLTW Biomedical Sciences Program (BMS) introduces high school students to the human body, cell biology, genetics, disease, and other biomedical topics in a sequence of four courses. The program prepares students for the postsecondary education and training necessary for success in a wide variety of positions, including physician, nurse, pharmaceutical researcher, and technician.[xx]

At the end of each PTE and BMS course, students complete an online course assessment, which enables teachers to track and document the development of critical thinking and problem solving skills as well as content and performance knowledge acquired. Students in certified schools who obtain high test scores are able to obtain college credit. Each year, PLTW schools administer grade-specific growth assessments in math and science to identify students' strengths and needs.

The successful local implementation of STEM initiatives—whether guided by PLTW or by other instructional frameworks—is highly dependent on a number of local factors that are different from community to community or school to school, including the availability of community and business partners and support from local two- and four-year colleges. A closer look at some innovative and mature STEM secondary school initiatives will provide a deeper picture of teaching, learning, and assessment experiences that effectively prepare students to pursue STEM education pathways in community and technical colleges.

STEM in Secondary School Settings: Some Exemplars

According to a recent National Academy of Engineering and National Research Council Report, STEM programs often neglect the relationships between STEM subjects:

> Most K–12 schools in the United States teach STEM subjects as separate disciplines, sometimes called "silos"—a math silo, a science silo, perhaps a technology education silo, and, in rare cases, an engineering silo—with few connections in curriculum, in teaching, and in classroom activities. Thus opportunities for leveraging the benefits of interconnections, such as using science inquiry to support learning of mathematical concepts, are largely lost. Students are left with an implicit message that each discipline stands on its own.[xxi]

Here, I describe several local STEM education initiatives that actively connect different STEM disciplines. Each of these high school initiatives illustrates a different kind of partnership with local colleges. These initiatives serve as potential models for replication and scaling up. They describe the vitality and creative strategies that high schools and community colleges must develop to ensure that the technical workforce needs of the twenty-first century are fully addressed.

- In Virginia, Thomas Jefferson High School for Science and Technology (TJHSST) offers a deep and rich academic curriculum that includes courses in organic chemistry, <u>neurobiology</u>, marine biology, <u>DNA</u> science, quantum mechanics, and computer science. All students complete <u>Calculus</u> (AB or BC) as well as a senior capstone project that entails either a yearlong research project or an off-campus mentorship through one of the school's <u>research labs</u>. Each of the thirteen labs allows for in-depth, multi-disciplinary research in a STEM field, including computer-aided design, optics and modern physics, automation and robotics, neuroscience, astronomy, and energy systems.

Representatives from business and industry and school staff collaborate in curriculum and facilities development. Recently, local business leaders and Jefferson parents established the Jefferson Partnership Fund to maintain and equip TJHSST's expensive labs.[xxii]

Over the past decade, fifteen states have established residential high schools serving talented high-school-age youth. Like TJHSST, many of these schools feature STEM-centric curricula or themes.[xxiii]

- In Iowa, Davenport West High School offers Project Lead the Way Engineering courses, including Introduction to Engineering Design, Principles of Engineering, Digital Electronics, Civil and Architectural Engineering, Computer Integrated Manufacturing, and Engineering Design and Development. Technology education teachers work closely with math and science teachers to integrate content and reinforce the knowledge and skills taught in each of the classes.

A large local partnership team meets regularly and provides funding and key support for a PLTW Summer Camp for eighth graders, day-long STEM awareness seminars, field trips, and scholarships to cover the college tuition fees for low-income students earning college credit for PLTW courses. The partnership team includes major employers as well as area colleges. The school has developed dual-credit articulation agreements with the University of Iowa, Iowa State University, and Eastern Iowa

Community College (EICC). Davenport's PLTW students who elect to concurrently enroll at EICC can also receive arts and science credits that transfers to four-year universities.[xxiv]

- In Wisconsin, TESLA Engineering Charter School offers a Project Lead the Way (PLTW) pre-college engineering curriculum within Appleton East High School. The school's operations are guided by a Charter Board composed of local business, college leaders, students, parents, and school administrators. In addition, Charter Board members and their organizations provide students with tour and job-shadow opportunities and support its Apple Corps Robotics Team.

A recent study compared the college and career readiness of TESLA seniors with other seniors at Appleton East.[xxv] On average, TESLA seniors:

✓ Received higher composite ACT scores (26.7 compared to 23.1)
✓ Received higher ACT math scores (27.1 compared to 23.2)
✓ Report greater involvement in career exploration, including talking with adults about career goals and participating in school experiences that help them clearly define careers goals.

These brief descriptions of promising high school STEM initiatives are just a small sample of the significant STEM innovation underway nationally and globally.[xxvi]

Fostering an Integrated STEM Education Pathway with Two-Year Colleges

High school STEM education initiatives have emerged rapidly over the past five years. However, as noted earlier, only 15 percent of the nation's high schools operate STEM-focused career academies, and only 12 percent of the class of 2004 had completed high school with a two-credit concentration in STEM-career-cluster courses (e.g., engineering technology, health sciences, or information technology). Sadly, less than one in five high school students are prepared to explore and pursue STEM-related careers and college majors.

Recently, the National Research Council's Board on Science and the Chief State School Officers acknowledged the importance of high school STEM education as a foundation for innovation and improving career and college readiness outcomes for all students. The new Framework for K–12 Science Education Standards outlines "significant improvements in how science is taught in the U.S." When this framework is translated into the Next Generation Science Standards by the Chief State School Officers and ACHIEVE, students will acquire a deeper knowledge of the practices of science (e.g., engineering, information technology applications, health diagnostics, and the like), which, in turn, will enable students to plan and carry out inquiries, compile evidence, and develop arguments. The Next Generation Science Standards will help educators and students acquire twenty-first-century answers to the age-old question: Why do I need to learn this?

The need to align and integrate science, engineering, and technology learning opportunities foreshadows other challenges that high schools must address. The National Academy's Engineering in K–12 Education Committee cited two major challenges in the evolution of STEM initiatives, both of which relate to the need for greater integration among STEM fields:

1. As STEM education is currently structured and implemented, it does not reflect the natural interconnectedness of the four STEM components in the real world of research and technology development.

2. There is considerable potential value, related to student motivation and achievement, in increasing the presence of technology, and especially engineering, in STEM education in the United States in ways that address the current lack of integration in STEM teaching and learning.[xxvii]

Additionally, the National Academy's Committee on Successful STEM Education came to a similar conclusion about the need for greater integration of STEM fields:

Research in technology and engineering education is less mature because those subjects are not as commonly taught in K–12 education. Although integrating STEM subjects is not the focus of this report, the committee recognizes the variety of conceptual connections among STEM subjects and the fact that science inquiry and engineering design provide opportunities for making STEM learning more concrete and relevant.[xxviii]

Beyond the lack of an integrated K–12 (and high school) STEM design, the practices this chapter has profiled reveal a number of other major challenges:

1. To date, the attention to four-year college STEM initiatives far outweighs the literature on two-year college STEM pathways. The vast majority of Math-Science Partnership grants, for example, have been awarded to partnerships involving four-year colleges and universities. The latest data on college-going patterns offer compelling evidence that more, not less attention should be given to two-year STEM pathways. Consider these data on college-going trends:[xxix]

   • Community colleges serve 44 percent of all undergraduate students in the United States and 40 percent of all first-time college freshmen.

   • Of the 12.4 million students enrolled in community colleges in 2008-09, 47 percent were 21 or younger, 58 percent were female, 39 percent were the first in their family to attend college, and 60 percent were enrolled in credit-based degree or certificate programs.

This profile of the rapidly increasing and diverse student population suggests that community college STEM programs can provide access to STEM employment opportunities to a broad sector of underrepresented students, including working adults, while simultaneously addressing the STEM middle-skill workforce-development needs outlined earlier. To encourage high school graduates and other young adults to enroll in two-year programs, high schools and STEM academies must develop adequate dual-credit STEM partnerships with community colleges.

2. Community and technical colleges must pay greater attention to outcomes for students pursuing Associate of Applied Science degrees. Most community colleges and their students have shifted away from transfer and general education missions toward preparing technicians. Now, preparing a skilled and technically competent regional workforce ranks as THE most

important community college mission. According to preliminary findings of the U.S. Department of Education's Committee on Measuring Student Success, the nation must create new student incentives and program-quality measures focused on learning and employment outcomes.[xxx]

3. Although federal funds have helped launch new high school STEM initiatives, in 2004 only one in five public high schools (21.5 percent) actually operated a career academy. Of these, only 6.4 percent offered a specialized career academy.

Prospects continue to improve for high school graduates who enter community and technical colleges with coherent STEM high school transcripts. According to a recent analysis of the class of 2004:

- 27 percent had enrolled in a two-year college by 2006.

- Students who had completed two- or three-credit STEM concentrations in high school (in engineering technology, health sciences, or information technology) were doing better on several measures than students who had completed concentrations in other fields. 87 percent of STEM concentrators were obtaining a credential or remaining enrolled in college, compared with only 82 percent of non-STEM students. 32 percent had enrolled in a college major that aligned with their high school concentration, compared with only 12 percent of non-STEM students. And 30 percent had obtained a first job or current job aligned with their high school concentration, compared with only 10 percent of non-STEM students.

- Among those who had completed a college prep program in high school (four credits in English, three in math, and three in science), about 15 percent had selected a STEM major in either a four-year or two-year college.

- Students who had completed both a college prep program of study and a two- or three-credit concentration in engineering or engineering technology were significantly more likely than other students to choose a STEM college major. Only a very small percent of students completed these concentrations, but of those who did, 51 percent of those in four-year colleges and 71 percent of those in two-year colleges chose a STEM major, compared to only 15 percent of students who completed only the college prep program.[xxxi]

Many communities do not have well-developed pathways to lead students from high school to STEM programs in community colleges. Even in communities that have strong STEM pathways, many students don't receive enough encouragement to consider pursuing them. For those that do, however, the future is indeed bright. As these findings suggest, it is imperative that we expand the number and quality of high school–community college STEM pathways.

Next Steps

For more than a century, America's community and technical colleges have provided fertile ground for preparing technicians and engineers. The important role of community colleges in educating engineers is not well known to the public, or even to the engineering community. In fact, 20 percent of engineering degree holders began their academic careers with at least ten credits from community colleges, and 40 percent of recipients of engineering Bachelor's and Master's degrees in 1999 and 2000 attended community colleges.[xxxii]
Community colleges offer unique opportunities for increasing diversity in the STEM workforce. The hallmarks of two-year colleges—access, affordability, transfer options, and flexible schedules—offer engagement strategies and incentives for low-income youth and adults, English-language learners, and other underrepresented populations.

Two key strategies can be developed in any community to enhance high school–community college STEM pathways.

First, close connections between high school and community college instructors will enhance students' postsecondary success. Together, high school and community college instructors can develop and deliver dual-credit courses in engineering, health sciences, and other technical fields. Many community colleges are experiencing rising remediation rates and low degree- or certificate-completion rates, and the rising concern for everyone–students, policymakers, and employers– is improving accountability and successful student outcomes. Toward that end, communities should adopt dual credit/dual enrollment programs like those included in many high school STEM initiatives. These programs are producing impressive results. For example, in New York and Florida, students who had earned dual credit in academic and career-technical courses prior to entering community colleges were significantly more likely to enroll full time, persist to the second semester and the second year, and possess a higher grade-point average.[xxxiii]

Second, the high-quality pedagogy used in many well-integrated high school STEM programs (e.g., integrated math/science/technical instruction, internships, industry mentors, student research projects, and formative or end of course assessments) provides a strong foundation that supports college success for all students. These integrated STEM-education approaches are very similar to the high-impact practices that students will encounter in two-year and four-year colleges, including learning communities, student-faculty research, service learning, and senior culminating experiences.[xxxiv] Well-developed high school STEM initiatives can help ensure the college success of all students.

[i] For details on the National Academy of Engineering's Grand Challenges for Engineering, see http://www.engineeringchallenges.org .

[ii] Career Academy Support Network, *Academies Nationwide*, http://casn.berkeley.edu/directories.php?us=1 (accessed September 5, 2011).

[iii] V. Bersudskaya and X. Chen, X, *Postsecondary and Labor Force Transitions Among Public High School Career and Technical Education Participants,* January 2011, http://nces.ed.gov/pubsearch/pubsinfo.asp?pubid=2011234 (accessed April 20, 2011).

[iv] James B. Conant, *The American High School Today: A First Report to Interested Citizens* (New York: McGraw-Hill, 1959).

[v] A. G. Powell, E. Farrar, and D. K. Cohen, *The shopping mall high school: Winners and losers in the educational marketplace* (Boston: Houghton Mifflin, 1985).

[vi] National Commission on Excellence in Education, *A Nation at Risk (Washington, DC: U.S. Department of Education, 1983).*

[vii] Committee on Prospering in the Global Economy of the 21st Century: An Agenda for American Science and Technology, National Academy of Sciences, National Academy of Engineering, Institute of Medicine, *Rising Above the Gathering Storm: Energizing and Employing America for a Brighter Economic Future* (The National Academies Press, 2007).

[viii] H. J. Holtzer and R. I. Lerman, *The Future of Middle Skill Jobs,* (Washington, D.C.: Brookings, February 2009), http://www.worksystems.org/portals/1/pdfs/career/Middleskilljobs09.pdf (accessed June 10, 2011).

[ix] A. P. Carnevale and S. J. Rose, *The Undereducated American* (Washington, D.C.: Center on Education and the Workforce, Georgetown University, 2011), http://www9.georgetown.edu/grad/gppi/hpi/cew/pdfs/undereducatedsummar y.pdf (accessed July 27, 2011).

[x] Missouri Department of Elementary and Secondary Education, "Perkins IV Programs of Study," http://dese.mo.gov/divcareered/perkins_iv_pos.htm (accessed July 16, 2011).

[xi] Association of Career and Technical Education, *Issue Brief: CTE's Role in Science, Technology, Engineering and Math*, June 2009, http://www.acteonline.org/uploadedFiles/Publications_and_Online_Media/fil es/STEM_Issue_Brief.pdf (accessed June 10, 2011), 1.

[xii] The STEM Academy, "Career Clusters and Programs of Study (Pathways) Mapped to the STEM Academy," http://www.stem101.org/pdf/perkins_application.pdf (accessed September 14, 2011 from http://www.stem101.org/grants.asp).

[xiii] National Science Foundation, Math and Science Partnership Program: Strengthening American by Advancing Academic Achievement in Mathematics and Science," http://www.nsf.gov/pubs/2010/nsf10046/nsf10046.pdf?WT.mc_id=USNSF_ 80 (accessed September 23, 2011).

[xiv] Abt Associates, Inc., *Government Performance and Results Act (GPRA) Indicators for the MSP Program, Performance Period 2008*, June 23, 2010 (accessed from http://www.ed-msp.net/index.php?option=com_content&view=article&id=5&Itemid=2 on September 15, 2011).

[xv] National Academy Foundation, "About NAF," http://naf.org/about-naf (accessed September 21, 2011).

[xvi] National Academy Foundation, "Our Themes," http://naf.org/our-themes (accessed September 15, 2011).

[xvii] National Academy Foundation, "Statistics and Research," http://naf.org/statistics-and-research (accessed September 15, 2011).

[xviii] Project Lead The Way, "Who We Are," http://www.pltw.org/about-us/who-we-are (accessed September 21, 2011).

[xix] Project Lead The Way, "Gateway to Technology" and "Pathway to Engineering," http://www.pltw.org/our-programs/engineering-curriculum (accessed September 15, 2011).

[xx] Project Lead The Way, "Biomedical Sciences," http://www.pltw.org/our-programs/biomedical-sciences-curriculum (accessed September 15, 2011).

[xxi] L. Katehi, G. Pearson, and M. Feder, *Engineering in K-12 education: Understanding the status and improving the prospects* (Washington, D.C.: National Academies Press, 2009), 20.

[xxii] Thomas Jefferson High School for Science and Technology, "An Overview," http://www.tjhsst.edu/discovery/labs/optics.php (accessed June 10, 2011).

[xxiii] R. F. Subotnik, R. H. Tai, R. Rickoff, and J. Almarode, "Specialized Public High Schools of Science, Mathematics, and Technology and the STEM Pipeline: What Do We Know Now and What Will We Know in Five Years?" *Roeper Review* 32, (2010): 7-16.

[xxiv] Project Lead The Way, "Davenport West High School," http://www.pltw.org/model-schools-2009-10-davenport-west-high-school (accessed July 17, 2011).

[xxv] L. A. Phelps, E. Camburn, and A. Lee, "High School STEM Engineering Programs of Study: A Value-Added Analysis (working paper, Center on Education and Work, University of Wisconsin, Madison, May 2011).

[xxvi] For more information, please see the following organizations' websites: TIES: Teaching Institute for Excellence in STEM, http://www.tiesteach.org/default.aspx; STEM Academy, http://www.stem101.org/index.asp; National Consortium of Specialized Schools of Mathematics, Science, and Technology, http://www.ncsssmst.org/; National Academies Press, http://www.nap.edu/topics.php?topic=282

[xxvii] See note 21 above.

[xxviii] Committee on Highly Successful Schools or Programs for K–12 STEM Education, *Successful K–12 STEM Education: Identifying Effective*

Approaches in Science, Technology, Engineering and Mathematics (Washington, D.C.: National Academies Press, 2011), 2.
[xxix] American Association of Community Colleges, *Fast Facts, 2011,* http://www.aacc.nche.edu/AboutCC/Pages/fastfacts.aspx (accessed July 27, 2011).
[xxx] See note 28 above.
[xxxi] L. A. Phelps, E. Camburn, and J. Durham, *Engineering the Math Performance Gap (*CEW Research Brief, Center on Education and Work, University of Wisconsin-Madison, June 2009), http://www.cew.wisc.edu/docs/resource_collections/CEW_PTLW_Brief_UWMadison.pdf (accessed June 10, 2011).
[xxxii] M. C. Mattis and J. Sislin, eds., *Enhancing the Community College Pathway to Engineering Careers* (Washington, D.C.: National Academies Press, 2005).
[xxxiii] M. M. Karp, J. C. Calcagno, K. L. Hughes, D. W. Jeong, and T. Bailey, *Dual Enrollment Students in Florida and New York City: Postsecondary Outcomes (*New York: Community College Research Center, Teachers College, February 2008).
[xxxiv] G. D. Kuh, *High-Impact Educational Practices: What They Are, Who Has Access to Them, and Why They Matter* (Washington, D.C.: Association of American Colleges and Universities, 2008), http://www.aacu.org/leap/hip.cfm (accessed March 27, 2011).

Creating Career Pathways for STEM Technicians

Dan Hull

For over 230 years, our great nation has shown—on many occasions and in varied ways—that its people and institutions can work together in innovation and diligence to create solutions that will assure *all its citizens* the right to enjoy equal opportunities to work and to prosper. We have been superior and exemplary in creating a democracy, defending our shores, caring for the aged and disadvantaged, and supporting public education and scientific research for health care and technology innovations. We can—and must—work aggressively to maintain our technical innovativeness, restore a career purpose to higher education, and provide appropriate educational opportunities for *all our youth* by recognizing and teaching to their interests, abilities, and learning styles.

The issues that have been set forth in the preceding chapters are these:

- Not enough engineering and science technicians are entering the workforce, leaving us without the technicians we need to ensure economic prosperity and secure national defense.

- Our public educational system is not adequately addressing the needs, interests, and learning abilities of "middle-quartiles" high school students, many of whom are our potential technicians.

- We already have the infrastructure (STEM High Schools) and the tools (ATE curricula for AAS degrees in emerging technologies) to solve these problems; we just need to mesh them together.

Simply stated:

There is a continuing inadequate supply of engineering and science technicians, primarily because too few high school graduates are either informed about or prepared to enroll in AAS degree programs at community and technical colleges. We must form articulated secondary/postsecondary partnerships to create a robust pipeline into these rewarding careers. These partnerships can expand and enhance high school STEM programs to create an appealing **curriculum option** *that will provide students with pathways to college and successful careers as technicians.*

These Career Pathways require a new curriculum strategy—one that will allow students who plan to enter AAS degree technician programs to enroll in appropriate courses during their junior and senior years of high school. This chapter describes a curriculum strategy and supporting elements that STEM high schools can use to provide these students with successful Career Pathways.

The Essential Elements for Preparing *All STEM Students*

STEM high schools should be recruiting and preparing capable students to enter postsecondary education programs to become scientists, engineers—*and technicians.* To serve all these students, high schools should incorporate the following curriculum and guidance recommendations throughout the secondary school grades.

- *Identify and encourage potential STEM students.*
  - ➢ Provide information, assessments, and experiences that will help students with the appropriate aptitudes and interests choose STEM education.
- *Encourage all students to develop a career interest and pursue their education with focus and purpose.*
  - ➢ Help students develop a career plan.
- *Provide the foundation for career success:*
  - ➢ Academic skills: The *base* upon which career knowledge and skills are formed. Includes appropriate proficiencies in mathematics, science, and communications.
  - ➢ Technical skills: A broad, interdisciplinary core, followed by instruction in the knowledge and skills peculiar to the

devices, materials, equipment, and processes of the chosen field. A flexible basis upon which to continue to learn new information as the technology inevitably advances.

➢ Soft skills: The ability to work with other people, serve on teams, solve problems, and be responsible for the accuracy and quality of one's work.

- *Validate the basis for STEM Curriculum Pathways:*

  ➢ Employer Requirements: Curricula and course content must be based on collective inputs from a broad and representative group of employers of these future workers. We must ask, "What do we want these graduates to know and be able to do when they go to work? What is their potential career ladder? Will they have the knowledge and skills necessary to advance in their field?" Course content should not be controlled by the unique interests and biases of faculty.

  ➢ Secondary/Postsecondary Coordination: One goal of secondary schools should be to prepare their graduates to be successful in the next level of their education. Secondary and postsecondary institutions should understand the potential and limitations of the curricula, cooperate to provide a seamless transition, encourage students to continue their educational and career pursuits, and eliminate duplication in course content from one level to another. Articulated, dual-credit courses give secondary students confidence to pursue postsecondary education, and a "head start" that can save them valuable time and educational expenses.

- *Ensure flexibility in student educational and career options.*

  ➢ Provide a broad curriculum base with options: Most students entering high school are not sure about what career they will want to pursue. Hopefully, they will have some interest in whether they will pursue a career in fields such as health care, science and technology, business, education, human services, hospitality, manufacturing, or others. This choice could place them in a career academy or a STEM high school. The curriculum in the 9th and 10th grades should be offered to all students in the academy, or specialty high school. Even if some students decide to change academies after the ninth grade, the broad base would provide them with ample credits to continue in another field.

➢ Offer alternative programs of study that can lead to either BS or associate degrees: During these early high school years, students will learn more about their interests, aptitudes, and career options. This will enable them to make choices in the eleventh and twelfth grades that would lead them toward postsecondary pursuits in either a baccalaureate degree program in science or engineering or an associate degree in a technical field.

➢ Moving toward specialization: In the eleventh grade, students could enroll in "technical core" courses that lead to careers in a broad cluster of technical specialties. For example, an electronics core could lead to a specialty in lasers, digital systems, control devices, instrumentation, robotics, communications, or other related fields. In the twelfth grade, students would choose a specialty and begin taking dual-credit courses that are offered in the freshman year of college. By slowly moving toward specialization, students are not "tracked" into a pathway that doesn't allow them to change their specialization without suffering a significant loss of credits—and time.

• *Foster student confirmation of a career pursuit. Encourage students to ask:*

➢ Is this really for me? Student experiences—including course content with applications, projects, job shadowing, and internships—are necessary to help students make good decisions.

➢ Students should also consider their interests and aptitudes and how those aptitudes and interests might guide them into baccalaureate or associate degree pathways. Students should be encouraged to ask themselves "What do I want to become?"

- ✓ A *scientist* who discovers and interprets new phenomena?
- ✓ An *engineer* who solves problems and designs new equipment and processes?
- ✓ A *technician* who makes equipment and processes work?

- *Provide experiences and discoveries to complement course work.*
  - ➤ Experiences with projects: Solving real work-related problems designed with input from employer representatives and evaluated by practicing workers.
  - ➤ Experiences with employers: Short-term, job-shadowing, part-time, and summer jobs within the environment of a student's chosen field.

- *Provide appropriate **student-oriented teaching.***
  - ➤ Maintain students' interest in course material by encouraging them to apply what they're learning and showing them how this knowledge is useful.
  - ➤ Teach scientific and technical content that is appropriate to different student learning styles. Some students are abstract learners and can master the required knowledge through class discussions and by reading traditional textbooks. Others are applied learners who benefit greatly from hands-on lab activities and problem-based teaching.

Creating Flexible STEM Curricula and Experiences: Including Technician Education by Building on the STEM Academy Model

All states have benchmarks for achievement that dictate standards for content and assessments to measure outcomes. Because standards and assessments vary significantly from state to state, it is not appropriate to propose a specific curriculum model for all schools. Instead, I offer more general recommendations for improving STEM education not only in high schools, but in earlier grades as well. I recommend that elementary and middle schools adopt the following practices.

The Early Grades

The primary functions of elementary and middle schools in STEM education are to:

- Cultivate students' interest in science and technology,
- Identify potential STEM students,
- Inform these students about STEM careers, and
- Encourage students to pursue STEM career preparation in high school.

Experts have identified many promising strategies for accomplishing these functions, including activities, tests, curricula and other experiences. Middle schools should expand or modify these strategies to ensure that they serve potential STEM technicians.

For example, identifying potential STEM students typically involves selecting seventh and eighth grade students who excel in math and science courses. But many potential STEM technicians may be in the middle quartiles of math and science achievement. These students are interested in science and technology and are "hands-on" or "applied" learners with strong spatial learning abilities. Educators can identify these students in two ways: by testing for high spatial learning abilities[i] and by observing students' performance in applied work. Potential STEM technicians often perform much better in labs and on projects than in classrooms or on textbook assignments. Many of these students have been "turned on" to science and technology through projects, experiences, and applied science courses.

In addition, when educators inform and encourage students about STEM careers, they need to make sure to discuss *STEM technicians*.[ii] The main idea to convey is that STEM technicians are the "geniuses of the labs" who have the unique ability and responsibility to "make things work." Teachers and counselors must also make sure that students know that they don't have to be an "A student" in science and math to become a technician; they just have to have a curiosity about how things work and the confidence to feel that "I can do it if I want to."

High School

Effective curricula for STEM programs in high schools have been tested and used extensively since 2005, and new developments and improvements surface frequently. These curricula typically involve a core that includes four-year sequences of mathematics, science, and English, in addition to other required courses in history, government, humanities and/or the arts.[iii] The math curriculum typically includes algebra, geometry, trigonometry, statistics, precalculus, and calculus. The science curriculum may include biology, chemistry, physics, Earth science, and environmental science.

Added to this core are special (elective) courses that introduce students to the basic tools and concepts of engineering and the biomedical sciences. Engineering courses may include engineering graphics and design, engineering strategies to problem solving, mechanical systems and design, biochemical systems, information systems, and digital electronics. Biomedical science courses may include human anatomy, biochemistry, and biomedical systems. These courses may be offered in the ninth, tenth, and eleventh grades. Advanced courses in engineering, biomedical science, and technology, offered in the eleventh and twelfth grades, are typically designed to lead students into BS programs at universities.

How the Typical High School STEM Curriculum Discourages Potential STEM Technician Students—And How This Can Be Altered

In the early high school years (ninth and tenth grades) two common practices can discourage potential technicians:

- Teaching abstract science and mathematics courses with few or no applications.
- Neglecting to teach about career opportunities for engineering, technology, and science technicians.

We have the resources we need to correct these practices, including curricula and teacher development programs that support application-oriented approaches to teaching mathematics and science. NSF/ATE centers and programs have developed career descriptions as well as descriptions of exemplary technicians in a variety of fields.[iv] By drawing on these and other resources, schools can adjust their practices to more effectively welcome and encourage middle-quartile achievers in the ninth and tenth grades to participate in STEM programs.

In the eleventh and twelfth grades, two additional practices practically eliminate potential technicians from STEM programs:

- Requiring STEM students to take advanced math courses, including Pre-Calculus and sometimes Calculus.
- Requiring STEM students to choose advanced courses designed to lead into BS science, engineering, and information technology programs.

The logical solution to this problem is to provide alternative eleventh- and twelfth-grade engineering technology curricula that will encourage and prepare potential STEM technicians to articulate into AAS-degree STEM programs at community and technical colleges— in other words, create *Career Pathways for STEM Technicians.*

The curricula for these programs of study are neither complicated nor expensive. They require three strategies:

- In the ninth and tenth grades, teaching Algebra and Geometry using problem-solving approaches and applications relevant to scientific and technical fields.
- In the eleventh and twelfth grades, offering a sequence of applied mathematics courses relevant to the technical courses taught in AAS-degree programs.
- Providing dual-credit technical courses that mirror those taught in the first year of the AAS-degree programs.

Chapters 6–13 will present examples of high school–college partnerships that have begun to implement these strategies.

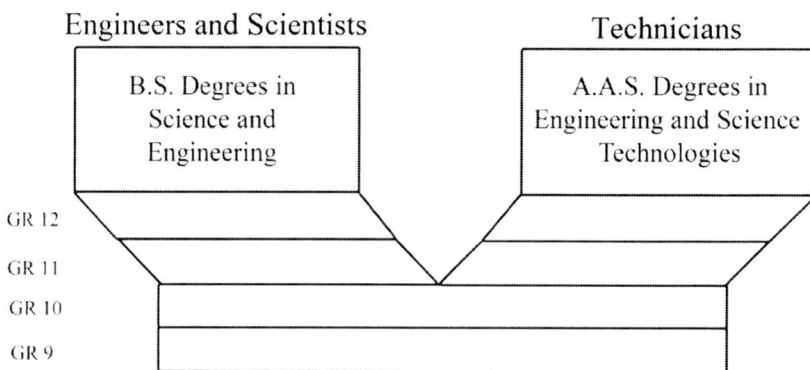

Figure 4.1. Alternative Curriculum Pathways for STEM High Schools

A Plan for a High School STEM Curriculum that Accommodates, Supports, and Encourages Potential Technician Students

1. Adjust the eleventh- and twelfth-grade math sequence.

The four-year math sequence in a typical STEM high school consists of Algebra I, Geometry, Algebra II, Pre-Calculus, and sometimes Calculus (for students who complete Algebra I in the eighth grade). An alternate "technician preparation curriculum" would diverge in the eleventh grade, at which point students would enroll in a problem-solving course. The problem-solving course would reinforce topics from Algebra I and Geometry, plus right angle trigonometry, that students will need to understand scientific and technical concepts in the AAS curriculum. These topics include:

Scientific Notation	Geometry
Powers and Roots	Special Graphs
Graphing in Rectangular Coordinates	Introductory Algebra
Trigonometry	Exponents & Logarithms
Complex Numbers	Angle Measure: 2 & 3 Dimensions
Unit Conversion	Sinusoidal Motion
Ratio and Proportion	

In the twelfth grade, students could take an applied version of Algebra II or Math Models and Applications, if such a course is required for high school graduation. A more useful twelfth-grade math course, however, would be a dual-credit college algebra course.

2. Create a sequence of technical courses that will prepare students to enter Engineering Technician AAS-degree programs.

Many high school STEM programs use the Project Lead The Way (PLTW) curriculum, described in chapter 3.[v]

The first tier of the PLTW High School Engineering Program, "Pathway to Engineering," consists of two or three courses taught sequentially in the ninth, tenth, and eleventh grades:

Introduction to Engineering Design (IED)

Designed for ninth- or tenth-grade students, the major focus of the IED course is to expose students to the design process, research and analysis, teamwork, communication methods, global and human impacts, engineering standards, and technical documentation. Students use 3D solid modeling design software to help them design solutions to solve proposed problems and learn how to document their work and communicate solutions to peers and members of the professional community.

Principles of Engineering (POE)

Designed for tenth- or eleventh-grade students, this survey course of engineering exposes students to major concepts they'll encounter in a postsecondary engineering course of study. Students employ engineering and scientific concepts in the solution of engineering design problems. They develop problem-solving skills and apply their knowledge of research and design to create solutions to various challenges, documenting their work and communicating solutions to peers and members of the professional community.

Digital Electronics (DE)—optional

Digital electronics is the foundation of all modern electronic devices such as cellular phones, MP3 players, laptop computers, digital cameras, and high-definition televisions. The major focus of the DE course is to expose students to the process of combinational and sequential logic design, teamwork, communication methods, engineering standards, and technical documentation. This course is designed for tenth- or eleventh-grade students.

The second tier of the PLTW Engineering program, typically offered in the eleventh and twelfth grades, offers two specialized courses that prepare students to enter one of the BS engineering disciplines.

While PLTW's first tier works well for all STEM students, its second tier does not effectively serve future technicians. An alternative eleventh- and twelfth-grade curriculum would include two to four technical courses taught in the first year of an AAS degree program at a nearby community or technical college. Technician students do not need to choose an AAS specialty area in the eleventh grade, because initial coursework supports multiple specialty areas. Usually, AAS degree programs are grouped into *clusters*. For example, an electronics cluster may support lasers, robotics, telecommunications, and instrumentation; a manufacturing cluster may support computer-aided design (CAD), automated system controls, or computer numerical control (CNC); and an alternative energy cluster may support solar and wind energy. If eleventh-grade students select a "clustered" course, they can then select a specialty area in the twelfth grade.

Dual-credit courses provide several advantages to both high schools and students.[vi] For high schools, dual-credit courses eliminate the need to design and develop new courses; they may also offer the opportunity to use the faculty and laboratories of the articulating college. Dual-credit courses offer students the opportunity to enter a postsecondary program while they are in high school, which reinforces their career opportunities and eases the anxiety of entering higher education. Most importantly, dual-credit courses provide students postsecondary credits that shorten the time required to earn an associate degree and save a considerable amount of money. Chapter 5 provides in-depth descriptions of dual-credit practices in technical courses and successful strategies for devising and reaching agreements about dual-credit courses.

Soph 2		Elective	Humanities		Technical Core	Technical Specialty	Technical Core
Soph 1	Elective		Social Science		Technical Core	Technical Specialty	Technical Core
Fresh 2	College Algebra			Physical Science	Technical Core	Technical Specialty	Technical Core
Fresh 1		College English		Physical Science	Technical Core	Technical Specialty	Technical Core
12th Grade	Algebra 2 w/Trig	English 12	Government	Physics	Health	Technical Specialty	Technical Specialty
11th Grade	Math Applications	English 11	American History	Chemistry	Physical Education	Technical Core	Technical Core
10th Grade	Geometry	English 10	World History	Biology	Physical Education	Foreign Language	Principles of Engineering
9th Grade	Algebra 1	English 9	Geography	General Science	Physical Education	Foreign Language	Intro to Engr Design

Figure 4.2. Suggested Secondary-Postsecondary Career Pathways for Engineering Technology (possible dual courses shaded)

Specialization	Core Courses	Specialty Courses
Photonics (lasers)	DC-AC Electricity Digital Electronics	Fundamentals of Light & Lasers Elements of Photonics
Biotechnology	Quality Assurance Solutions & Biotech Assays	Introductory Biotechnology Biomanufacturing
Advanced Manufacturing	Properties of Materials Digital Electronics	Manufacturing Processes Control Systems
Communications Systems	DC-AC Electricity Introduction to TCP/IP	Communication Systems Unified Networks (Data, Voice and Video Services)
Information Technology	Computer Hardware and Operating Systems Computer Programming	Introduction to TCP/IP Programming for Mobile Devices
Microsystems (MEMS)	DC-AC Electricity Digital Electronics	Introduction to MEMS MEMS Fabrication, Design and Integration
Nanotechnology	Reactions, Forces and Interactions Computer Modeling at the NanoScale	NanoMaterials Nanoscience Applications

Figure 4.3. Examples of Technical Core/Technical Specialty Courses for Eleventh and Twelfth Grades

These courses have been developed for freshman-level, postsecondary AAS degree programs, and are appropriate for dual-credit use in STEM high schools.

Making the Changes Necessary to Accommodate Potential STEM Technicians

Change is never easy, particularly in public education. Providing an alternate pathway in STEM high schools will require selection and approval of different courses and may require significant professional development for teachers and counselors. But the major barrier to making the changes recommended in this chapter is the perception that the only road to career and lifestyle success is the baccalaureate degree. Parents, students, teachers, counselors, and some school administration and governing boards hold this faulty perception, and correcting it will require information, insight, and advice from employers. Employers must be organized and encouraged to inform the public of the potential, need, compensation, and career paths for workers with postsecondary education at the associate-degree and certificate levels.

The other barrier that must be overcome is the bias against students in the middle quartiles of achievement, including the "not my children" attitude of many parents. We must believe that these students are capable and deserving of an equitable, alternate path to career and life success. We simply must rid our schools and communities of the "silent agreement" that education for middle-quartile students should be second rate or nonexistent. An alternate educational pathway and different teaching styles for these students are not only deserved and useful, but also absolutely vital to the future of our country.

---

[i] Later chapters will discuss testing for spatial learning ability.

[ii] Later chapters will elaborate on the kinds of work that STEM technicians do.

[iii] See chapter 3 for more detail about STEM high school curricula.

[iv] See the following chapters and http://www.atecenters.org for more information about technician careers in a variety of fields.

[v] For more information, please see http://www.pltw.org.

[vi] A later chapter will discuss dual-credit courses in more detail.

Dual Enrollment's Role in STEM Pathways

Katherine Hughes
Institution on Education and the Economy
Community College Research Center at Teachers College, Columbia University

There is a great deal of change currently underway in our nation's K–12 and higher education systems. There is general agreement that some education after high school is a requirement of today's world, but while most students pursue college, many never complete any college credential. The national effort to implement Common Core State Standards is a far-reaching attempt to increase rigor and provide all students with the skills they need to be successful as they transition to postsecondary study and training. At the same time, many colleges and college systems are reviewing their curricula and programs, particularly in developmental education, to try to address low graduation rates.

As with reform movements in the past, these efforts are not as integrated as would be ideal. In the previous chapter, Dan Hull argues, as many have before him, that secondary and postsecondary institutions should be strongly cooperating to provide students with a seamless transition. This chapter will describe how dual enrollment—permitting high school students to take college courses—fosters collaboration across the education sectors to help students progress more smoothly and quickly along their pathway. An early start in college can also help students seriously consider their interests and aptitudes, another important point Hull raises, to answer the important question "Is this really for me?"

Dual Enrollment/Dual Credit Definitions and Prevalence

In dual enrollment, high school students take college courses and, if they pass them, earn college credits. Frequently, when state or local policy allows, students can earn both high school and college credit for the same course, which is referred to as dual credit. Unfortunately, terminology is not standard across the nation—"concurrent enrollment" is another term frequently used, usually to refer to a particular program format in which the college courses are offered on the high school campus and taught by qualified high school teachers.[i] "Joint enrollment" is used in Iowa to refer to high school students enrolled in community college courses.

Dual enrollment is a means of earning early college credit that is different from participation in Advanced Placement (AP) or International Baccalaureate (IB). In dual enrollment, high school students take actual college courses, while in the other programs, students take high school courses that have rigorous, college-level content. In AP and IB, whether students earn college credit is dependent on their score on a single, end-of-course examination and whether the college they enter will give credit for that particular score. In contrast, dually enrolled students complete whatever assignments and assessments would normally be given as part of the college course, and are given a final grade on a college transcript that looks like any regular college student's transcript.

Articulation, traditionally a part of Tech Prep programs, is another potential means of college credit-earning for high school students. Tech Prep was originally devised in the 1980s as a sub-baccalaureate college pathway for students that would provide marketable skills in technical fields. The program, which has been federally funded through the Carl D. Perkins Act, promotes the concept of articulation through close collaboration among high schools and colleges—connecting course sequences and content across high school and college occupational programs to reduce duplication and provide a transparent pathway for students.

However, the articulation model has suffered from being locally based and institution-specific. In many cases, the credits have been awarded only after students matriculated to the partnering college, and sometimes only if they declared their major in the same field as their

Tech Prep courses. An evaluation of Tech Prep found that many student participants were unaware that they could earn college credits from their high school Tech Prep coursework.[ii] Thus, as programs attempt to improve access to and portability of college credits earned in high school, we are seeing a movement away from articulated courses and toward incorporating dual enrollment into Tech Prep and other college transition programs.

Due in large part to the focus on "college for all" in this country, dual enrollment has increasingly been encouraged as part of a broad college transition strategy. U.S. Department of Education national surveys found that, during the 2002–03 school year, 71 percent of high schools reported that they had students in dual credit courses, with over 800,000 students nationwide participating.[iii] While the national surveys did not track student participation over time, several states are doing so, and providing evidence of a dramatic increase. For example, in Iowa, 2010 saw a record high of 38,283 students in dual enrollment, up 14 percent from the previous year. Jointly enrolled students in that state now account for about a quarter of total community college enrollment.[iv] In addition, states such as Florida, Texas, and New Mexico have passed new policies promoting dual enrollment participation.

A career/technical education (CTE) focus is evident in many dual enrollment programs and state policies. The national surveys found that just over one-third of the dual credit courses taken in the 2002–03 school year had a career or technical focus.[v] In Georgia, from 2001 to 2004, enrollment of high school students in technical college courses increased 93 percent, compared to a 7 percent increase in the number of high school students in the general population.[vi] Iowa has also been a leader in career-focused dual enrollment through its career academies, programs that link the last two years of high school with a postsecondary career preparation program. The state provides supplemental funding to school districts for dually enrolled high school students, thus encouraging participation.

As dual enrollment has expanded, a range of formats and models have sprung up. For some time around the country, qualified high school students have had the opportunity to independently enroll in college courses on college campuses. Now, dual-enrollment programs are increasingly operated on the high school campus with qualified high school teachers serving as instructors. Another model is the early or middle college program, in which a high school may be located on a college campus and students are expected to earn a year or more of college credit.[vii] Some courses mix high school and regular college students, while some contain high school students only. And courses may be offered before, during, or after the regular school day.[viii]

The Evidence on Dual Enrollment

As noted above, student participation in dual enrollment has grown tremendously over the past decade or more. This growth began to occur even before there was any evidence of the effectiveness of dual enrollment. A 2003 review of the literature found mostly descriptive reports on dual enrollment and almost no evidence that dual enrollment contributed to students' college access or academic success.[ix]

Why has there been such enthusiasm for the strategy? Dual enrollment has existed for many years, but until the last decade, it typically targeted more advanced students—usually those who had taken all the advanced courses their high schools offered and thus were looking to neighboring colleges for additional courses. As the college-for-all notion began to gain national support, policymakers, educators, and foundations began to argue that a broader range of students could benefit from participation—that dual enrollment could serve as a college-readiness strategy. The argument was that early exposure to college while in the supportive high school environment could help to prepare students for subsequent college success. Dual enrollment provides participating students with the opportunity to improve their academic and nonacademic skills, to understand what will be required of them in college, and to gain motivation for future college attendance by realizing that they are indeed capable of doing college work. The experience has been described by one researcher as a "role rehearsal" in which students "try on" the role of a college student. [x]

Attention also began to be paid to dual enrollment opportunities as part of career-focused pathways in high schools. There is a good foundation of evidence that high school career and technical education plays a role in reducing dropout rates and increasing high school graduation rates, and that "well-designed career-focused programs can improve employment, earnings, non-academic skills, and career choices, particularly for at-risk and low-income youth."[xi] A career focus can engage students through applied learning and help them see pathways through college to future employment. Thus, it is increasingly common for dual enrollment to be one element of a career-focused program such as a Program of Study (the next generation of Tech Prep) or a Career Academy (called Partnership Academies in California). These and other programs include academic and career/technical content and aim to help students meet college- and career-readiness standards.

Several studies have now been completed that show a range of benefits for dual-enrollment students. In general, researchers have found that earning college credits prior to high school graduation reduces time to degree and increases the likelihood of earning a college degree.[xii] Studies on specific dual-enrollment programs have found participation to be associated with a range of positive high school and college outcomes, including college enrollment and persistence. For example, studies conducted by the Community College Research Center (CCRC) of students in Florida and in New York City found that participation in dual enrollment was positively related to students' likelihood of earning a high school diploma, to college enrollment, to persistence in college, and to higher postsecondary grade point averages.[xiii] Both studies examined effects for students concentrating in career/technical fields in particular (the Florida sample included all students as well as a CTE sub-sample; the New York City student sample consisted of only those in career/technical high schools). The benefits were present for both CTE and non-CTE students, and, importantly, the benefits to dual enrollment were stronger for male and low-income students, both of which are subgroups of concern today. Additional research on the City University of New York (CUNY) College Now program, the same program studied by CCRC, also found positive results.[xiv]

Recent findings from two studies by the National Center for Postsecondary Research (NCPR) suggest that where students take dual-enrollment classes and what classes they take are important.[xv] In one study, also using data from Florida, NCPR found that dual-enrollment participation had positive effects on college enrollment and degree attainment for students who took classes on a college campus but no effects for those who took the college classes at their high school. In the second study, which used a different methodology,[xvi] no benefits were found across the entire sample of student participants, but large positive effects were found for students who took the college algebra course.

Since these are not randomized controlled trials, one cannot say for sure that the positive outcomes are not due to student motivation or other pre-existing student characteristics. In addition, some researchers have brought to light concerns associated with dual enrollment, such as limited oversight, uncertain academic quality or rigor (referring to courses held on high school campuses), and program costs.[xvii] Still, there is widespread support for the idea of a blending of high school and college courses in students' high school senior year that would allow for some early accumulation of college credits. And, the growing research base on dual enrollment provides support for broadening access and in particular for encouraging participation by students in career/technical programs, males, and disadvantaged students.

Building a Strong Partnership and Program

Central to dual enrollment is the partnering of secondary and postsecondary institutions. The education literature is replete with reports bemoaning the misalignment between our K–12 and postsecondary education sectors. In dual enrollment, the line between the two becomes more malleable as students, staff, and faculty move and work across it. The following are some steps and guiding questions that can help educators move toward a strong dual-enrollment partnership.[xviii]

Step 1: Creating a Partnership
High schools should investigate whether there are multiple potential partner colleges in the vicinity, or one obvious choice. High schools should examine whether the college provides courses, programs, and degrees in career fields related to courses offered by the high school. Both high schools and colleges should build on any existing relationships and take advantage of any college outreach or partnership office that already exists. High schools and colleges need to determine

the division of roles and responsibilities between the institutions, as well as day-to-day oversight and high-level support.

Step 2: Understanding State and District Regulations

Most states have policies pertaining to dual enrollment; these usually address how programs may be funded and which students are eligible to participate.[xix] Districts, counties, and postsecondary institutions may impose additional regulations. It is critical to understand these regulations and the limitations or opportunities they present. For instance, some public school districts do not grant high school credit for college courses, which prevents schools from offering dual enrollment classes during the school day and can discourage enrollment.

Step 3: Choosing Dual-Enrollment Courses

Course selection can influence student and program success. Student population and ability should be carefully considered when choosing courses. Colleges might require students to pass placement tests to register in some courses. Technical and more hands-on courses are good at engaging students in the course and the career field. Courses that engage students on a personal level, such as student success courses that introduce them to college and help them plan their education and career trajectories, can also be very successful. Courses should be presented in clearly recommended sequences.

Step 4: Determining Course Logistics: Where, When, Who?

As noted above, dual-enrollment programs are offered today in many different formats. Courses may be offered on the college or high school campus and may be taught by college instructors or high school teachers who qualify as adjuncts. Students tend to perceive classes held on the college campus as more authentic. However, if the college is far away, the difficulty of getting there might lead to low enrollment or persistence rates. If classes are held on the high school campus, special attention must be paid to quality and rigor.[xx] Opening high-school-based dual enrollment courses to college students can help ensure authenticity. Instructor quality is also important; dual enrollment instructors should be committed to teaching the target population and should use pedagogies that engage students.

Step 5: Offering Supplemental Supports

In many cases, support services are critical in addressing high school students' academic and college knowledge needs and in monitoring student progress. Clear and frequent communication between course instructors and high school staff is important (staff may need to obtain written consent from participating students to access their college grades). Tutoring, supplemental instruction, and outside-class activities are often a part of dual enrollment programs. To coordinate and manage student support services, many programs have found that it is helpful to assign or hire a dedicated staff member.

Step 6: Determining Costs and Sources of Funding

Staff should determine program costs and consider possible sources of support. Costs include those for instruction, books, transportation (when necessary), staff, student recruitment, support services, and data collection. Funding for dual enrollment tends to come from state or local sources, foundation and other private sources, and tuition or fees.[xxi] Some states allow both high school districts and postsecondary institutions to claim per-student apportionment for dually enrolled students; such an arrangement provides an incentive for both institutions to participate. But not all states provide such support; in some cases, state policy directs school districts to pay funds to the higher education institution to cover part of the costs. State and local policies also vary with regard to tuition and fees; some states direct or allow colleges to waive such costs for students.

Step 7: Promoting and Sustaining Dual-Enrollment Programs

Recruiting students for dual-enrollment programs can sometimes be challenging, given the many other activities high school students are involved in. Both high school and college staff must be enlisted in this effort and must work together to develop attractive recruitment messages and materials. Some programs send representatives to high school parent nights to publicize dual-enrollment opportunities. To ensure that programs have sufficient institutional support, staff should work to develop relationships with higher-level administrators at colleges and school districts.

Step 8: Measuring Success

Evaluation is essential for understanding and improving program implementation and outcomes. Data should be collected on participating student characteristics and outcomes, including early accumulation of college credits, a remediation-free start to college, persistence in college, and credential completion.

In addition to desired student outcomes, there are other, harder-to-measure yet still positive outcomes for the institutions involved. Dual enrollment can help school personnel from both sectors become better at their jobs. Working with high schools and teaching high school students can help college instructors better understand the academic challenges their regular college students struggle with. In turn, high school teachers may learn how unprepared their students are for college work, which may prompt conversations with their college counterparts about the standards they should be aiming for.

Dual Enrollment for STEM Technician Pathways

In chapter 4, Dan Hull clearly lays out the sequence of courses and content that would prepare and encourage students to pursue STEM technician careers. In contrast to the Project Lead the Way program, which offers eleventh- and twelfth-grade courses that purposefully lead into baccalaureate engineering programs, Hull proposes that students take college technical courses during those years. Below are some examples of programs pursuing this model.

Abraham Lincoln High School Biotechnology Academy, San Francisco, California

This academy was founded in 1995 and has grown since then from one section of thirty students to five sections and over 150 students. With hundreds of biotech firms within commuting distance of San Francisco, there are many opportunities for employment at various levels of skill. The academy has a close relationship with City College of San Francisco, which offers several certificates in the field: a biomanufacturing certificate, biotechnician certificate, and a stem cell research certificate. Several years ago, the college began to offer academy students a college-credit stem-cell culturing class that includes both lecture and lab components. Students gain practical experience with the laboratory equipment and gain knowledge and skills to advance in this field.

EE-Tech, Redding, California
The Emerging Energy Technologies Program is sponsored by the Shasta Union High School District in partnership with Shasta Community College. The intent of the program is to fill a need in the growing green sector by providing students with training and course pathways in the renewable energy (wind and solar) industry, while simultaneously promoting college attendance. Two courses from the college's Industrial Technology Core—Industrial Trades Basics and Electricity and Electronics—are offered on the high school campus. Students earn high school and college credits for these courses, and can go on to complete the core requirements and continue in their choice of technology pathway, for example, photovoltaic or wind wind-generation studies. An important aspect of this program is the professional development provided to high school teachers in the construction trades areas, to help them integrate renewable energy technologies into their courses and thus cultivate student interest in the field.

Conclusion

Allowing high school students to explore and pursue their varied interests is an important part of helping them discover their talents and settle on a career path. While there is general agreement that some education after high school is a requirement of today's world, all students will not pursue a single path to and through postsecondary education. Even the Common Core State Standards movement, which aims for consistency in K–12 student learning outcomes, assumes flexibility:

> One of the hallmarks of the Common Core State Standards for Mathematics is the specification of content that all students must study in order to be college and career ready. This 'college and career ready line' is a minimum for all students. However, this does not mean that all students should progress uniformly to that goal.[xxii]

Dual enrollment is an important means of providing students opportunities to take interesting and challenging courses that build on their high school coursework. It also stimulates secondary and postsecondary educators to collaborate and align their curricula. Such efforts can only help students be better prepared for and better understand the possible next steps in their lives.

---

[i] See the National Alliance of Concurrent Enrollment Partnerships: http://www.nacep.org.

[ii] Debra D. Bragg. *Promising Outcomes for Tech-Prep Participants in Eight Local Consortia: A Summary of Initial Results* (St. Paul, MN: National Research Center for Career and Technical Education, 2001).

[iii] Tiffany Waits, J. Carl Setzer, and Laurie Lewis, *Dual Credit and Exam-Based Courses in U.S. Public High Schools: 2002–3*, NCES 2005-009, U.S. Department of Education (Washington, DC: National Center for Education Statistics, 2005); Brian Kleiner and Laurie Lewis, *Dual Enrollment of High School Students at Postsecondary Institutions*: 2002–3, NCES 2005-008, U.S. Department of Education (Washington, DC: National Center for Education Statistics, 2005).

[iv] State of Iowa, Department of Education, *Joint Enrollment Report 2010* (Des Moines, IA: Community Colleges and Workforce Preparation, 2010).

[v] Waits, Setzer and Lewis, *Dual Credit.*

[vi] Richard Lynch and Frieda Hill, "Dual Enrollment in Georgia's High Schools and Technical Colleges," *Techniques: Connecting Education and Careers* 83, no. 7 (October 2008): 28-31.

[vii] E. Barnett, C. Hindo, K. Buccerri, and J. Kim, J. Ten *Key Decisions in Creating Early Colleges: Design Options Based on Research* (New York: National Center for Restructuring Education, Schools, and Teaching, Teachers College, Columbia University, 2011).

[viii] For a review of various program formats and the implications the formats have for including disadvantaged students, see Linsey Edwards, Katherine L. Hughes, and Alan Weisberg, *Different Approaches to Dual Enrollment: Understanding Program Features and Their Implications* (New York: Community College Research Center, 2011).

[ix] Thomas Bailey and Melinda Mechur Karp, *Promoting College Access and Success: A Review of Dual Credit and Other High School/College Transition Programs* (Washington, DC: U.S. Department of Education, 2003).

[x] Melinda Mechur Karp, "Facing the Future: Identity Development among College Now Students" (Ph.D. dissertation, Columbia University, 2006).

[xi] Richard Kazis, "What We Know about Secondary Schools and Programs that Link School and Work," in *Remaking Career and Technical Education for the 21st Century: What Role for High School Programs?*, ed. Richard Kazis (Boston, MA: Jobs for the Future, 2005).

[xii] Clifford Adelman, *The Toolbox Revisited: Paths to Degree Completion from High School through College,* (Washington, DC: U.S. Department of Education, 2006); Joni Swanson, *An Analysis of the Impact of High School Dual Enrollment Course Participation on Postsecondary Academic Success, Persistence, and Degree Completion* (Iowa City, IA: Institute for Research and Policy Acceleration at the Belin-Blank Center for Gifted Education, University of Iowa, 2008).

[xiii] Melinda Mechur Karp, Juan Carlos Calcagno, Katherine L. Hughes, Dong Wook Jeong, and Thomas R. Bailey, *The Postsecondary Achievement of Participants in Dual Enrollment: An Analysis of Student Outcomes in Two States* (St. Paul, MN: National Research Center for Career and Technical Education (NRCCTE), University of Minnesota, 2007).

[xiv] Sam Michalowski, *Positive Effects Associated with College Now Participation for Students from New York City High Schools: Fall 2003 First-Time Freshman Cohort* (New York: CUNY Collaborative Programs, Office of Academic Affairs, 2007).

[xv] Ceclia Speroni, *High School Dual Enrollment Programs: Are We Fast-Tracking Students Too Fast? An NCPR Working Paper,* (New York: Teachers College, Columbia University, National Center for Postsecondary Research, 2011); Cecilia Speroni, *Determinants of Students' Success: The Role of Advanced Placement and Dual Enrollment Programs, An NCPR Working Paper* (New York: Teachers College, Columbia University, National Center for Postsecondary Research, 2011).

[xvi] This study used a regression discontinuity design. NCPR compared outcomes for Florida students scoring just above and below the minimum GPA required by the state to qualify for dual enrollment. The students who scored above the GPA cutoff were able to participate in dual enrollment, while those who scored just below the GPA cutoff — and who were statistically indistinguishable from their barely-passing peers in characteristics known to affect outcomes — could not.

[xvii] Drew Allan, *Dual Enrollment: A Comprehensive Literature Review and Bibliography* (New York: The City University of New York, Office of

Academic Affairs, Collaborative Programs, 2010). Also see Lorna Collier, "College Credit for Writing in High School: Bypassing First-Year Comp Courses Is a National Trend with Both Critics and Defenders," *The Council Chronicle*, March 2011, 11-13.

[xviii] These steps were adapted from Linsey Edwards and Katherine Hughes, *Dual Enrollment for High School Students* (New York: Community College Research Center and Career Academy Support Network, 2011).

[xix] Katherine L. Hughes, Melinda Mechur Karp, Baranda Fermin, and Thomas Bailey, *Update to State Dual Enrollment Policies: Addressing Access and Quality* (Washington, DC: U.S. Department of Education, Office of Vocational and Adult Education, 2005).

[xx] The National Alliance of Concurrent Enrollment Partnerships has determined quality standards for high-school-based dual enrollment programs and serves as a national accrediting body.

[xxi] Michael Griffith, *Funding Dual Credit Programs: What Do We Know? What Should We Know?* (Denver, CO: Education Commission of the States, 2009).

[xxii] Common Sore Standards Initiative, *Common Core State Standards for Mathematics, Appendix A*, http://www.corestandards.org/assets/CCSSI_Mathematics_Appendix A.pdf.

Secondary/Postsecondary Programs of Study For Photonics Technicians

John Souders, National Center for Optics and Photonics
Education (OP-TEC)
Greg Kepner, Indian Hills Community College

What is Photonics?

Photonics is one of the fastest growing technology areas in the world. In formal terms, it is defined as *the technology of generating and harnessing light and other forms of radiant energy whose quantum unit is the photon.*

In less scientific and more technology-based terms, photonics can also be defined as *the study and design of devices and systems such as lasers and optical fibers that depend on the transmission, modulation, or amplification of light.*

Though both definitions are accurate, they are somewhat esoteric. At the National Center for Optics and Photonics Education (OP-TEC), we have simplified these definitions and describe photonics as:

> *The technology that generates and uses light to enhance the quality of our lives.*

The main source for generating light in photonics is the laser. All of us have heard about lasers and know from movies like *Star Wars* and other science-fiction productions that they have some very interesting capabilities. It is a lesser-known fact that lasers, fiber optics, and other photonics devices have substantially increased the quality and significantly lowered the costs of many products we use. Bar code scanners, which use lasers and light sensors, are in most retail stores and have revolutionized the processes of customer checkout and inventory management. We listen to high quality music on inexpensive CD players whose main component is a laser. Lasers have given us a level of health care that was unattainable in years past.

Laser light carries millions of phone calls along fiber-optics lines, providing clear, undistorted voice signals. Every time we make a copy or print out a document, photonics technology is being used. Applications of photonics technology have grown nearly exponentially since the first laser was produced in the early sixties. Light will always be at the center of technology expansion. Since the speed of light is the speed limit of the universe, photonics will continue to be a part of any new technologies that spring forward.

The future of photonics is very secure, and engineers, scientists, and technicians who work in it will have substantial opportunities to advance professionally and live high-quality lives. High school students who are trying to select a STEM career pathway can be assured that photonics is here to stay and will continue to provide ample opportunity for challenging, rewarding, and stable careers. This is especially true for photonics technicians because the industry demand for them far exceeds the supply.

Photonics Technician Demand and Supply

In 2012, The National Center for Optics and Photonics Education (OP-TEC) commissioned the University of North Texas Survey Research Center (UNTSRC) to determine the demand for photonics technicians within the United States. To accomplish this task, UNTSRC contacted by telephone a large random sample of industrial companies throughout the United States that conduct research and development (R&D) with optics, lasers, and photonics technology; are original equipment manufacturers (OEM) for R&D companies; or use laser equipment. The goal was to determine how many photonics technicians the companies employed, the additional number they would need in 2012, and the number they would need through 2016. The researchers identified a total of 4,217 U.S. photonics companies and contacted over 2,500 of those companies before generating a representative sample of 629 companies, of which 345 employed photonics technicians. Besides national employment data, the survey collected similar data for seven different regions in country. After performing a statistical analysis of the survey, the researchers drafted a final report.[i] The results showed that over 19,000 photonics technicians are employed in the United States, that over 1,590 additional photonics technicians are needed in 2012, and that approximately 630 more will be needed each year through 2016.

To determine our capacity to produce an adequate number of photonics technicians, it is necessary to not only look at the demand side of the equation, but also the supply side. In 2008, OP-TEC also commissioned a study to determine the capacity of U.S. colleges to meet the national demand for photonics technicians.[ii] The study identified all two-year colleges in the United States that have active photonics instructional offerings (at least one course in photonics, optics, and/or laser technology) and have former students employed in the optics/photonics industry as a result of their participation in that instruction. The results of the survey indicated that approximately 750 students are enrolled in programs that are leading to technician-level employment in the photonics industry. However, only about 250 students are completing their programs each year and are available for employment as photonics technicians.

With a demand over the next several years of over six hundred photonics technicians per year and U.S. two-year colleges only able to supply approximately 250 completers, there exist strong employment opportunities for people with photonics skills. This obvious shortage has led employers to try to find "work-arounds." For instance, employers are providing photonics training to technicians in their organizations whose educational backgrounds are in other fields, hiring engineers to fill technician jobs, and providing high school graduates on-the-job training. Though these measures do add to our supply of technicians, employers have made it very clear that these stopgap measures are far more costly—in time and money—and produce a lower quality workforce than if they were able to fill their open photonics technician positions with people with AAS degrees. The bottom line is that a high demand will exist into the foreseeable future for people with AAS degrees in photonics. With average starting salaries for photonics technicians at $40,000,[iii] the future is certainly bright for those students who choose this career path.

OK, the future is bright—lots of job opportunities and good starting salaries—but what kind of work do these technicians do that puts them in such great demand and earns them these kinds of salaries?

Functions and Skills of Photonics Technicians

Skill standards are employer specifications for the knowledge and skills required for success in specified technical areas. OP-TEC has engaged employers from around the country to develop a set of skill standards specifically for photonics technicians. Figure 6.1 provides detail on the structure, organization, and components of these standards. Copies can be downloaded from the OP-TEC website (http://www.op-tec.org).

Photonics Skill Standards	Organization of the Standards

1. Specifies the knowledge and skill requirements for a variety of technicians in the photonics industry.
2. Provides the foundation for AAS curriculum and materials development in photonics technology.
3. Can be adapted by local employers for curriculum design at a particular college.
4. Provides a benchmark (4+2) curriculum framework, an infusion curriculum, and an advanced certificate.
5. Identifies six specialty areas for photonics technicians.
	• Six specialty areas: *Communications, Lighting and Illumination, Medicine, Maufacturing, Optoelectronics, Imaging, and Remote Sensing*
	• Critical work functions for each specialty
	• Tasks
	• Skills
	   – Employability
	   – Technical

Critical Work Functions General areas of responsibility or functions required of a technician working in a specialty area. *Example:* Assemble various fiber-optic components and modules into subsystems and understand their function.	**Tasks** Observable and measurable activities that technicians perform to accomplish a critical work function. *Example:* Integrate fiber-optic components and modules into specified systems.
Technical and Employability Skills Basic abilities that are necessary for a technician to perform a task. *Examples:* • Technical: Test and verify initial source output and launch angles at source/fiber interface. • Employability: Navigate the Internet to gather task-related information.	

Figure. 6.1. Details on the structure, organization, and components of OP-TEC's *National Photonics Skill Standards for Technicians*

OP-TEC's *National Photonics Skill Standards for Technicians* was designed to assist two-year colleges in developing their AAS photonics curricula and also to provide curriculum guidance in modifying existing AAS programs to add photonics courses. This latter case is becoming very prevalent, especially in technical programs where photonics is, or is becoming, an enabling technology.

The OP-TEC standards are very detailed and cover the six photonics areas listed in figure 6.1. It is beyond the scope of this chapter to present, for all six areas, the various functions a photonics technician must perform and the associated skills they must possess. However, a careful review of the six areas shows that there are many common functions and skills that technicians in each area must possess to meet workplace demands. For instance, technicians in all six areas must be able to build, test, modify, install, operate, calibrate, maintain, and repair laser and electro-optics devices and systems. In other words, photonics technicians must understand and have hands-on experience in all aspects of a variety of lasers, components, detectors, and related equipment in fields that use photonics as an enabling technology.

This review also reveals that to accomplish these functions, photonics technicians must be able to:

- Use the basic principles, concepts, and laws of physics and optics in practical applications.
- Use algebra and trigonometry as problem-solving tools.

And they must be able to build their education and training on this knowledge to:

- Analyze, troubleshoot, and repair systems.
- Use materials, processes, procedures, equipment, methods, and techniques common in photonics.
- Apply detailed knowledge in photonics with an understanding of applications and industrial processes in that field.
- Use computers for information management, equipment and process control, and design.
- Record, analyze, interpret, synthesize, and transmit facts and ideas with objectivity.
- Communicate information effectively by oral, written, and graphical means.

From this review of the photonics skill standards, we can readily see that photonics technicians are very hands-on workers who are adept at applying math and science concepts to solve technology-related problems. As mentioned in the chapter 4, these technicians are the "geniuses of the lab" and masters of "making things work." It is for these reasons they are in such demand and command good salaries.

But how do photonics technicians gain this knowledge and these skills? In the next section, we will describe a curriculum model that will answer this question.

A Secondary/Postsecondary Career Pathway for Photonics

Chapter 4 presented a template for a STEM curriculum to serve as a career pathway for technical fields based on engineering or science. Since photonics is this type of technical field, we were able to use this template to generate a model STEM curriculum for photonics technicians. This curriculum is presented in figure 6.2.

Soph 2		Elective	Humanities		Laser Devices	Laser Electronics	Laser Measurements
Soph 1	Elective		Social Science		Trouble Shooting and Repair Techniques	Laser Technology	Laser Components
Fresh 2	College Algebra			Physical Science	Computer Aided Design	Geometric and Wave Optics	Programmable Logic Controllers
Fresh 1		College English		Physical Science	Analog Devices	Introduction to Lasers	Electronic Devices
12th Grade	Algebra 2 w/Trig	English 12	Government	Physics	Health	Elements of Photonics	Circuit Analysis
11th Grade	Math Applications	English 11	American History	Chemistry	Physical Education	Fundamentals of Light and Lasers	Digital Electronics
10th Grade	Geometry	English 10	World History	Biology	Physical Education	Foreign Language	Principles of Engineering
9th Grade	Algebra 1	English 9	Geography	General Science	Physical Education	Foreign Language	Intro to Engr Design

Figure 6.2. Model Curriculum for Photonics Technicians
(Shaded Courses are Dual-Credit)

The course titles used in this curriculum were chosen to represent topics that appear in typical photonics AAS programs built from an electronic core curriculum.[iv] This is a "model curriculum"—a starting point from which high schools and two-year colleges can begin working together to build a photonics career pathway for their students that meets state educational requirements and satisfies the skill standards required by local employers.

As pointed out in chapter 4, one way of building a robust student pipeline through a career pathway is to provide within a curriculum opportunities for students to take dual-credit courses. The photonics model curriculum offers this opportunity. The three specialty courses (AC/DC, Fundamentals of Light and Laser, Elements of Photonics) are standard in AAS photonics curricula and would earn students college credit upon successful completion. Since these courses are so rich in technology applications, math and science simulations, computer integration, and laboratory and problem solving experiences, it should not be difficult to find high school courses with similar content that could be adjusted to merit dual credit. The three Project Lead the Way Tier 1 courses that are part of this model curriculum are also strong candidates for dual credit. One of OP-TEC's partner colleges, Iowa's Indian Hills Community College, is currently offering college credit for all three of these courses. If all six courses (AC/DC, Fundamentals of Light and Lasers, Elements of Photonics, and the three Project Lead the Way courses) are dual credit, students could earn more than eighteen college credits prior to leaving high school. Depending on the capability of the student, there is also a potential within the model photonics curriculum for students to earn dual credit for mathematics, science, and English courses as well. The financial and motivational implications of accelerating a student's education in this way are obvious.

Though building a standards-based, state approved curriculum with dual-credit opportunities is a viable means for attracting students to a photonics technician career pathway, the curriculum, by itself, is not enough to ensure that the student pipeline will stay full and flowing. "Pipeline building" must also include making students aware of career opportunities in photonics and motivating them to seek ways to become a part of this field.

As chapter 4 indicated, generating awareness about technical careers can begin as early as the elementary grades. For example, photonics professionals could visit classrooms, provide students exciting demonstrations of photonics concepts and applications, and share information on what photonics technicians do and what educational preparation they need. For some students, an awareness of photonics will spark an interest in it. Effective pipeline building provides students a means to nurture this interest and determine if photonics would be a good career choice for them.

We have found at OP-TEC that Summer Institutes held on two-year college campuses are very effective at providing this nurturing[v] These institutes are typically a week in duration and provide high school students opportunities to work in photonics labs and learn about careers in photonics. Through the labs, students get to see firsthand the excitement and challenges of working in the photonics area, and for many students, this where an interest in the field starts. The institutes also introduce students to college life on campus, providing them a much better base from which to make the decision to pursue a college degree.

Another pipeline-building strategy that OP-TEC has tested is the use of a *focused, high school recruiter*. Most colleges have recruiters that attend career and educational fairs at high schools. Typically, these recruiters represent all technical programs taught at their colleges. Their understanding of any one particular program is usually, and understandably, limited. OP-TEC has supported several colleges in assigning dedicated recruiters to their photonics programs.[vi] As the name implies, the recruiter's sole responsibility is to recruit students into their college's photonics program. The college's photonics faculty members train the recruiter and prepare him or her to answer detailed technical questions as well as questions about their program's curriculum and career opportunities in the field. These recruiters are relatively young professionals with a marketing background and effective people skills. They make regular visits to high schools, give presentations on photonics, and connect interested students with faculty at the college. Their compensation is often tied to recruiting goals set by the college's photonics department. Colleges that have used these dedicated recruiters have seen positive gains in their photonics program enrollments. If a high school has in place a STEM

Photonics Technician Career Pathway, a major goal for both the summer institutes and dedicated recruiters would be to recruit students onto this pathway by the time they enter eleventh grade.[vii]

The following photonics program description shows a curriculum with all the basic features of the model curriculum, but with additional dual-credit courses.

Indian Hill Community College's Early College Program

Indian Hills Community College (IHCC) is one of OP-TEC's Partner Colleges. Its Laser/Electro-Optics program is one of the largest in the country. IHCC photonics graduates are sought by employers throughout the country. In 2011, IHCC received over ninety job opportunities for its eighteen graduates. We have selected IHCC's Early College Program to showcase not only because it is highly successful and fully implemented, but also because it demonstrates how a STEM Technician Career Pathway can be implemented at the high school level. The curriculum used in this program is an adaption of the model photonics curriculum that has been adjusted to meet the needs of employers. It is a testimony to what can be achieved when educators and employers join forces to provide students opportunities for rewarding careers.

In 2007, IHCC identified a severe national shortage of skilled workers in the advanced manufacturing industry, which is highly dependent on the use of lasers for operations such as welding, cutting, grinding, and precision measurements. Since IHCC had an AAS program in Laser Electro-Optics and Robotics/Automation—both essential to advanced manufacturing—it was well positioned to help remedy this labor shortage. IHCC leaders felt that this shortage could be rectified most rapidly by decreasing the amount of time it takes for students to complete the AAS degree. Their solution was to develop and implement an early college program.

The IHCC Early College program allows participating high school students to start AAS programs in their junior year of high school and, in a three-year period, graduate with an AAS degree in Laser/Electro-Optics, Electronics Engineering, or Robotics. Without this early college start, students would have to complete their junior and senior years of high school before entering IHCC, and then spend two years earning an AAS degree. Instead of graduating in three years, a high school junior would take four years to earn an AAS degree. Thus, the early college program places high school juniors into the workplace one year sooner than a traditional program would. Reducing the time it takes students to become work-ready has the effect of increasing the number of workers available to an industry, which is exactly what the advanced manufacturing industry needs to rectify its shortage of qualified technicians.

The Early College program closely follows the model photonics curriculum presented earlier in this chapter. This curriculum is shown in figure 6.3.

Fresh 4					Optical Systems Analysis	Photonics Systems Lab	Photonics Applications
Fresh 3		Comm Elective			Physical Optics	Photonics Repair	Automated Laser Processing
Fresh 2					Geometric Optics	Laser System Fundamentals	Optical Devices
Fresh 1				Science Elective	Intro to Solidworks	Introduction to Photonics	Laser Components
12th Grade	College Algebra	English Elective/ Workplace Comm	Ethics/ Gov't	Physics	Intro to Computers Physical Education	Photonics Concepts Power Transfer Technology	Digital Electronics
11th Grade	Technical Math	English 11	American History	Chemistry	Physical Education	AC/DC Circuit Analysis	Analog Devices Business Essentials
10th Grade	Geometry	English 10	World History	Biology	Physical Education	Foreign Language	Principles of Engr
9th Grade	Algebra 1	English 9	Geography	General Science	Physical Education	Computer Applications	Intro to Engr Design

Figure 6.3. IHCC Early College Photonics Curriculum
(Shaded Courses are Dual-Credit)

The IHCC Early College curriculum has some unique features not seen in the other presented curriculum models. For instance, since IHCC uses a quarter system, the postsecondary portion of the Early College curriculum is shown as four distinct quarters (Fresh 1—Fresh 4). In addition, this curriculum includes only one year of postsecondary coursework. This is because one full year of courses that are normally taught at the postsecondary level are now taught as dual-credit offerings in the junior and senior years of high school. Eliminating the need for a second postsecondary year allows students in the Early College program to reduce by one year the time it would normally take to enter the workforce with an AAS in Electro-Optics; as stated previously, this acceleration was one of the main goals of this program. This curriculum also features split boxes that result from the difference in course length of high school and postsecondary courses. Typically, a high school course lasts a full academic year, which is equivalent to two quarters at IHCC.

The IHCC curriculum is an adaptation of the model photonics curriculum presented in figure 2. Both offer three Project Lead the Way courses in their high school offerings, and both provide high school students opportunities to complete college-level technical core and specialty courses in photonics. The major adaptation IHCC made to the model photonics curriculum is the addition of two technical core courses to the high school portion of the curriculum. These additional courses were needed to meet IHCC's goal of reducing the amount of time it took students to earn their AAS in Laser Electro-optics. This adaptation is an example of how secondary and postsecondary educators can start with the model photonics curriculum and modify it to meet their local and regional needs.

Implementation of this early college curriculum required close coordination between IHCC and high school faculty and administrators. Students take the state-mandated high school courses at their local high schools (which can include the ninth- and tenth-grade tier 1 Project Lead the Way Courses in the model photonics curriculum), but from 8:00 to 10:50 each morning, they take courses at IHCC. These courses, which constitute the Early College program, include Project Lead the Way Digital Electronics, DC Circuit Analysis, AC Circuit Analysis, Analog Circuits, Photonics Concepts, and related mathematics and communications courses. Students who participate in the Early College program in both their junior and senior years complete all requirements in IHCC's electronics core and leave high school with an Electronics Technician diploma. With the electronics core completed, students then enter IHCC full time in the summer term and begin work on any of three technology programs: Laser/Electro-Optics, Electronics Engineering, or Robotics/Automation. Participants who enter the Early College program as juniors are able to complete an AAS degree in one of the three program areas one year after high school graduation. IHCC also gives college credit for the ninth- and tenth-grade Tier 1 Project Lead the Way Courses included in the model photonics curriculum. Adding the credit students earn from completing IHCC's electronics core to the college credit awarded for these other two Project Lead the Way courses, students in the Early College program can graduate from high school with forty-five college credits.

During the first year of the program in 2007, seven high school seniors from three different high schools enrolled in the program. Six of those students went on to earn Associate Degrees at IHCC. Of those six students, one student went on to work in retail sales, one student pursued an Associate of Arts degree, one student graduated with an AAS degree in Robotics/Automation Technology, and four students graduated with AAS degrees in Laser/Electro-Optics Technology. The Laser and Robotics graduates began careers in their fields of study with pay that averaged $40,000 per year.

IHCC is continuing to move forward with dual enrollment programs. The Early College program served as a pilot program that is now fully implemented. IHCC is attempting to meet the needs of area high school students who have STEM career goals but wish to pursue these goals as AAS technicians rather than as BS-level scientists or engineers.[viii]

Final Thoughts

Before we conclude this chapter, it is important to note that what photonics technicians bring to the workplace is both important and unique. The education of photonics engineers and scientists is not focused on the hands-on, equipment-oriented types of knowledge and skills that a photonics technician possesses. Though scientists and engineers can create experiments and design innovative ways to apply technology, it takes the knowledge and skills of a technician to ensure the required data is collected and recorded and that designs become working models or prototypes. The knowledge and skills of photonics technicians supplement and complement those of engineers and scientists and are essential to achieving success in any technology-based effort. Though our society places greater value on the contributions of scientists and engineers, much of their efforts would come to naught without the efforts of technicians. A STEM Technician Career Pathway that starts in high school and ends with an AAS degree is vital to the viability of our country's photonics industry and will provide a means of reducing the erosive economic effects of outsourcing high-tech jobs to foreign competitors.

[i] Darrell M. Hull, Paul Ruggiere, and Paul Illich, *Photonics Technician Employment in the United States: An Industry Survey of Current and Future Demand in 2012 for Education and Training Programs* (Waco, TX: OP-TEC Monograph, forthcoming in summer 2012).

[ii] Darrell M. Hull and Robert S. Gutzwiller, "Two-Year College Enrollment and Completion Rates: An Estimation of Workforce Supply for Technicians in Optics, Photonics, and Laser Technology" (OP-TEC Working Paper Series, Working Paper #5, 2008).

[iii] Hull, Ruggiere, and Illich, *Photonics Technician Employment.*

[iv] Dan Hull and John Souders, *Providing Photonics Education for Technicians* (Waco, TX: OP-TEC Monograph, April 2010); John Souders and Dan Hull, *A General Curriculum Framework for Infusing "Enabling" Technologies into Postsecondary Technical Programs* (Waco, TX: OP-TEC, April 2010).

[v] Feng Zhou, *Outreach Activities to Enlist High School Students for Electro-Optics Technician Programs at Indiana University of Pennsylvania, Northpointe Two-Year Campus (Waco, TX:* OP-TEC Monograph, May 2009); Larry Grulick and John Pedrotti, *TSTC Waco's Photonics Summer Institutes for High School Science and Technology Teachers* (Waco, TX: OP-TEC Monograph, May 2009).

[vi] Chrys Panayiotou, *Transforming Electronics Engineering Technology by Infusing Photonics* (Waco, TX: OP-TEC Monograph, May 2008).

[vii] More details about the summer institutes and dedicated recruiters can be found at the OP-TEC web page (http://www.op-tec.org).

[viii] For more information on IHCC's Early College program, contact Greg Kepner (greg.kepner@indianhills,edu).

All references can be downloaded at the OP-TEC website (http://www.op-tec.org).

Secondary/Postsecondary Programs of Study for Biomanufacturing Technicians

Sonia Wallman
Northeast Biomanufacturing Center and Collaborative (NBC²)
Montgomery County Community College, PA

What is Biomanufacturing?

In its simplest terms, biomanufacturing is a form of manufacturing that uses the machinery of living cells to produce a product of interest. Living cells make all sorts of products: specialized gland cells in spiders make silk, mammary gland cells make milk, yeast cells turn grapes into wine and grain into beer.

An understanding of modern biomanufacturing begins with a discussion of biotechnology. Biotechnology is a field of applied biology that involves the use of living organisms and bioprocesses in engineering, technology, medicine, and other fields requiring bioproducts. Biotechnology also utilizes these products for manufacturing purposes; we call this *biomanufacturing*. Modern applications of biotechnology include genetic engineering as well as cell- and tissue-culture technologies. The concept encompasses a wide range of procedures for modifying living organisms according to human purposes. The history of biotechnology goes back to domesticating animals, cultivating plants, and "improving" animal and plant life through breeding programs that employ artificial selection and hybridization.[i]

Biomanufacturing engineers and technicians have learned to make silk in sheep mammary gland cells. They have also learned to make other useful products, such as biopharmaceuticals, biofuels, bioplastics, and industrial enzymes. Biomanufacturing engineers can even construct new organs: with regenerative medicine techniques, a few cells can be plucked from one's body in a cleanroom atmosphere and grown into a new "replacement" organ.

Because the cells used to biomanufacture products are so small, many of them are needed in order to make enough product to sell commercially. To accomplish this, engineers "scale up" the production process: they use very large containers, or "bioreactors," to grow the cells, and they use piping to transfer cells and product for further processing. Automated process control is introduced to control the cells' and products' environment and movement through the biomanufacturing facility.

Biopharmaceutical manufacturing is the most highly regulated biomanufacturing industry because most of its products are delivered within the body via the blood stream; thus, they have to be pure and meet high quality control standards. The United States Food and Drug Administration (FDA) ensures that biopharmaceutical manufacturing proceeds according to current Good Manufacturing Practices (cGMP). Individuals with education and training in biomanufacturing at community colleges usually enter the field as entry-level technicians. If they are interested, they can pursue further education that will open up many career pathways within the industry. Though not really difficult to learn, the field requires from three to six months of hands-on training to develop proficiency.

Tasks Performed by Biomanufacturing Technicians:
- Demonstrate an understanding of the biotechnology manufacturing process.
- Adhere to the documentation guidelines of current Good Manufacturing Practices (cGMP)
- Execute a wide variety of laboratory techniques in microbiology, biochemistry, and molecular genetics, including (but not limited to) solution preparation, gene cloning, DNA extraction and amplification, library construction, hybridization, forensic analysis, cell culture, and protein production, purification, and verification.
- Generate and maintain accurate documentation, including laboratory notebooks, batch records, and log books.

- Analyze and draw conclusions from generated scientific data, and present findings.
- Use critical thinking and principles of logic to analyze ethical issues raised in the practice of biotechnology.

Because the public knows relatively little about career paths for biomanufacturing technicians, the Northeast Biomanufacturing Center and Collaborative (NBC^2) conducts *Protein is Cash* workshops for local high school teachers at community college biotechnology laboratories.[ii] The purpose of these workshops is to aid in awareness of the career paths available and to help in the development of a local source of biomanufacturing technicians to fuel the local and national bioeconomies. NBC^2 has worked with the United States Department of Labor to develop four registered biomanufacturing apprenticeship certificates for four biomanufacturing occupations. Certificates in upstream processing and downstream processing qualify students to work in production, and certificates in microbiology and biochemistry qualify students to work in quality control. The registered biomanufacturing certificate indicates that a student has received on-the-job training in biomanufacturing and an associate's degree in biotechnology/biomanufacturing.[iii]

Biomanufacturing Technician Demand and Supply

Industrial biomanufacturing—whether it be making beer, yogurt, biopharmaceuticals, biofuels, or industrial enzymes—is a growing industry that is fueling a brand new bioeconomy. Like other sectors of manufacturing, the jobs in biomanufacturing require skills in production, quality control, and regulatory compliance, as well as specialty areas associated with production, including process development, facilities, metrology, validation, and environmental health and safety.

Biomanufactured products are made of or by living cells. Facilties technicians provide clean steam for sterilizing bioreactors, water for injection (WFI) for growing the cells, and clean-room air. Metrology technicians maintain and calibrate instrumentation. Validation technicians validate that equipment and processes will result in a product that meets previously delineated standards for purity, identity, strength, or activity. Production technicians must nurture the cellular factories that make the bioproduct (upstream processing) and then purify these cellular products (downstream processing). Quality control technicians sample upstream and downstream processes to ensure that the product meets previously delineated standards. And process development technicians develop the scale-up process for the bioproduct. Biomanufactured products include biopharmaceuticals, replacement organs, nutriceuticals, industrial enzymes, and biofuels, to name a few.

Biopharmaceutical manufacturing represents the maturation of the biotechnology industry. Biotechnology is a new, growth industry that began in 1982 with the commercial manufacture of human insulin in *Escherichia coli* by Eli Lilly and Company in Indianapolis, Indiana. Today, there are 320 large biopharmaceutical manufacturing facilities in the nation and many, many more mid- and small-sized operations.[iv]

Biopharmaceuticals are one of the most complex commercial products to manufacture. Though it is automated to a degree, it still takes many people to make a single biopharmaceutical: most facilities employ between 100 to 499 people. Typically, 30–35 percent of biopharmaceutical employees work in production, including both upstream processing or cell culture, where the cells make the product, and downstream processing or purification, where the product is purified away from everything else. Another 13 percent of biopharmaceutical employees typically work in quality control, using microbiology or biochemistry to test the product for purity, identity, strength, and activity.

Today's biopharmaceutical manufacturing is a highly people-oriented business. Hiring, training, and retention, especially in production areas, are critical to the success of the industry, locally, nationally, and globally. According to 2011 data, hiring and retention rank very highly on the list of capacity constraint issues.[v]

Employment in the U.S. bioscience sector reached 1.42 million in 2008, a gain of 19,000 bioscience industry jobs since 2007. The biosciences continued to grow during the first year of the recession. At the height of the recession, our nation's bioscience manufacturers could not find qualified workers to manufacture biopharmaceuticals and other bioscience products.[vi] In 2011, the nation's biopharmaceutical manufacturers brought on board a variety of new staff: 33.2 percent in production and 13 percent in regulatory, quality-assurance, or quality-control areas. These numbers are expected to rise by 2015.[vii]

Biopharmaceutical manufacturing appears to be the vanguard for biomanufacturing many other products in crossover "green" industries, such as the biomanufacture of biofuels to replace our current reliance on fossil fuels. It also appears to be a gateway for many other bioproducts, including replacement organs, nutriceuticals, biofuels, bioplastics, and industrial enzymes.

There is no doubt that the demand for biomanufacturing technicians will continue to grow. A 2009 survey of twenty-three biomanufacturers indicated that they expected to hire two hundred biomanufacturing technicians in 2011 and approximately 840 biomanufacturing technicians in 2013.

There are 272 degree programs in biotechnology at two- and four-year colleges throughout the nation. More than 50 percent of these programs are at community colleges, and fifty of these degree programs have a biomanufacturing course or component. In fall 2009, 2,700 students were enrolled in these programs at forty-three institutions. There were 461 completers from college and 61 from high school programs in fall 2010; 64 percent entered the workforce, and 41 percent matriculated at universities.

Functions and Skills of Biomanufacturing Technicians

Skill standards are employer specifications for the knowledge and skills required for success in specified technical areas. NBC^2 has engaged employers from the United States and Ireland to develop a set of skill standards specifically for biopharmaceutical manufacturing technicians. Figure 7.1 provides details on the structure, organization, and components of these standards. Copies can be downloaded from the NBC^2 website at http://www.biomanufacturing.org/bioskillstandards.html.

Biopharmaceutical Manufacturing Skill Standards	Organization of Standards
1. Specifies the knowledge and skill requirements for a variety of technician career paths in the biopharmaceutical manufacturing industry. 2. Provides the foundation for A.S., A.A.S, and Certificate curriculum and materials development in biomanufacturing. 3. Can be used for additional curriculum design and development. 4. Identifies ten specialty areas for biopharmaceutical manufacturing technicians: upstream processing, downstream processing, quality control microbiology, quality control biochemistry, quality assurance, process development, validation, maintenance and instrumentation, and environmental health and safety. 5. Much of the knowledge and skills gleaned learning biopharmaceutical manufacturing can be used in the biomanufacture of biofuels, industrial enzymes, nutriceuticals, bioplastics, and replacement organs.	• <u>Ten specialty areas</u> with a focus on production and quality control. These focal areas include upstream processing or cell culture, downstream processing or purification, quality control microbiology, and quality control biochemistry. Other areas include quality assurance, facilities, metrology, validation, environmental health and safety, and process development. • Critical work functions for each specialty • Tasks • Skills -Employability -Technical
Critical Work Functions General areas of responsibility or functions required of a technician working in a specialty area. *Example:* Perform upstream manufacturing operations.	**Tasks** Observable and measurable activities that technicians perform to accomplish a critical work function. *Example:* Monitor cell concentration by cell counting or measuring OD.
Technical and Employability Skills Basic abilities that are necessary for a technician to perform a task. *Examples:* • *Technical:* Familiarity with analytical instrumentation. • *Employability:* Understanding the bioproduct development process from discovery research through process development and manufacture for commercial production.	

Figure 7.1. Details on the structure, organization, and components of NBC[2]'s *Biopharmaceutical Manufacturing Industry Skill Standards for Technicians*[viii]

NBC[2] has used the biomanufacturing skill standards in the development of a textbook— (*Introduction to Biomanufacturing*), a hands-on laboratory manual for college students, and a *Protein is Cash* laboratory manual for high school teachers and students. The *Protein is Cash* manual comes with online support materials, including virtual laboratory-scale and industrial-scale modules that can be used for practicing and assessing hands-on skills in production, as well as PowerPoint presentations and scientific articles to support both knowledge and skills training.[ix] The biomanufacturing skill standards have been used in developing the registered United States Department of Labor biomanufacturing apprenticeships and will be used to develop a biomanufacturing technician exam and certification.

The NBC^2 Biomanufacturing Skill Standards are detailed and cover the ten specialty areas listed in figure 7.1. It is beyond the scope of this chapter to present, for all ten areas, the key functions that biomanufacturing technicians must fulfill and the associated skills they must possess. However, a careful review of the ten areas shows that there are many common functions and skills that technicians in each area must possess to meet workplace demands. For instance, technicians in all areas must be able to understand basic troubleshooting and problem solving tools and methodologies including failure modes and effects analysis (FMEA), fault tree analysis, and root cause analysis.

The areas of commonality for biopharmaceutical biomanufacturing technicians can be seen in the *Competency Crosswalk*, in which 80 percent of the following critical work functions are shared amongst six biomanufacturing technician jobs (upstream and downstream processing, quality control microbiology and biochemistry, facilities, and metrology):[x]

- computer use
- career skills
- critical thinking
- regulatory, safety, and environmental compliance
- general industry knowledge
- aseptic processing

Many critical work functions of the following areas of work are more specific to that particular career pathway: facilities, metrology, production (upstream and downstream processing) and quality control (microbiology and biochemistry).

From this review of biopharmaceutical manufacturing skill standards, one can see that biomanufacturing technicians are very hands-on workers who are able to use their biomanufacturing skills and knowledge to make a desirable bioproduct.

The skills, knowledge, and attributes needed to manufacture biopharmaceuticals can be applied to crossover biomanufacturing industries such as the biomanufacture of biofuels, industrial enzymes, bioplastics, and replacement organs, in which:

1. Cells are used to scale up a product of interest, which is then purified to certain agreed-upon characteristics (production).
2. Sampling is done during the process of manufacturing the product of interest, and data is collected to show that the product of interest has the agreed upon characteristics (quality control).
3. Equipment and processes are validated and operated according to standard operating procedures (SOPs), and production is documented through the use of batch records (quality assurance).
4. The biomanufacturing facility is computer controlled.
5. The product is marketed and sold to customers. (Pursuing a biomanufacturing career pathway would ensure that you would be competitive on the job market; you may even be able to obtain a sales position for a biomanufacturer.)

But how do biopharmaceutical biomanufacturing technicians gain this knowledge and these skills? In the next section, some curriculum models will answer this question.

A Secondary/Postsecondary Career Pathway for Biomanufacturing

The following STEM curricula for high school and community college biotechnology programs provide early entry into career pathways for biomanufacturing technicians and a background that is transferable to four-year colleges and universities. Dual credit is common in science and mathematics courses.

Soph 2			Biochemistry	Bioethics	Biomanufacturing	Internal or External Internship or Apprenticeship
Soph 1			General Chemistry 2	Humanities/ Foreign Language/ Fine Arts	Discovery Research	Principles of Genetics
Fresh 2	Writing Technical Documents	Probability and Statistics	General Chemistry 1	Social Sciences	Microbiology	
Fresh 1	College English	College Algebra	General Biology		Introduction to Biotechnology	Information Technology
12th Grade	English 4	Algebra 2 w/Trig	Physics	Government	Health	Internal or External Internship
11th Grade	English 3	Math Applications	Chemistry	American History	Physical Education	Microbiology or Introduction to Biotechnology
10th Grade	English 2	Geometry	Biology	World History	Physical Education	Foreign Language 2
9th Grade	English 1	Algebra I	General Science	Geography	Physical Education	Foreign Language 1

Figure 7.2. Model High School and Associate Degree Curriculum for Biomanufacturing Technicians
(Shaded courses are possibly dual-credit.)

Attracting High School Students to Biomanufacturing Career Pathways

The Biomanufacturing Apprenticeship Program

The Biomanufacturing Apprenticeship program was started at Great Bay Community College (formerly New Hampshire Community Technical College) in 2004 in collaboration with the U.S. Department of Labor and the DOL-funded National Centers for the Biotechnology Workforce.[xi] This unique program saves students money and allows them to pursue an early academic and industry biomanufacturing track that leads to further education and a rewarding career. The stories of four apprentices follow:

Katrice

Katrice attended a high school biotechnology program and graduated in June 2005. She entered a community college biotechnology/biomanufacuring program and was the very first biomanufacturing apprentice at a nearby biomanufacturer. Katrice graduated in May 2005 and was offered a full-time position at the same company where she still is employed. In 2009, she took advantage of the company's free tuition benefit and began a bachelor's degree program in microbiology at a nearby university, where she received full credit for her two-year associate degree in biotechnology. At the age of twenty-four, Katrice just bought her own home. Katrice is featured in a video that the Department of Labor produced to commemorate the seventieth anniversary of its Registered Apprenticeship system.[xii]

Joey

In his junior year of high school, Joey completed a high school internship under the guidance of a community college faculty member. He decided to enroll in that community college after graduation, and in the summer after his first year at community college, he began a biomanufacturing apprenticeship at a nearby biomanufacturer. He graduated early, after the fall semester of his second year at the community college, because he had sixteen dual credits in science and mathematics. He finished his Associate in Science degree in three instead of four semesters. After graduating in December, he was employed at his apprenticeship site until he began a bachelor's degree program in the fall. He is pursuing a career pathway in biomanufacturing process development. You can hear Joey's story in the career video *What is Biomanufacturing?*, produced by Advanced Technological Education Television (ATETV).[xiii]

Shain

Shain graduated from high school in 2006. He greatly enjoyed learning about science, but he was unsure how to make a career out of it. As Shain told ATETV, "I was very interested in my high school's chemistry, physics, and biology classes, but I was really unsure where to continue from there." Shain found out about the apprenticeship program at a nearby community college, which sold him on joining the biotechnology/biomanufacturing program there. "It's great—most of my high-school friends are still working the same retail jobs, and I work in a growing high-tech industry! The apprenticeship program opened doors for me that wouldn't have been there otherwise. Biopharmaceutical manufacturing is a multi-billion dollar business here in New England, and it's changing our lives. I love being a part of it all." After graduating with an A.S. degree, Shain transferred his credits to a nearby university and graduated with a bachelor's degree in microbiology in May 2011. He is now employed as an upstream processing technician in a nearby biopharmaceutical manufacturing facility. A video documenting Shain's story can be seen on ATETV's *Community College Programs.*[xiv]

David

Before David graduated from high school, he took a course in biotechnology through a dual-credit program. He learned to prepare slides, use a microscope, inoculate plates, perform electrophoresis tests, and perform more high-tech procedures. Once he learned about the advanced, hands-on facilities in use at a nearby community college, he became interested in enrolling. Then he learned about the Biomanufacturing Apprenticeship Program, and this made the community college opportunity even more appealing. David gained experience in his apprenticeship and focused his skills and goals early. "I'm interested in genetics and genetic engineering," says David, who after graduating with an A.S. degree took a technician position at Millennium Pharmaceuticals in Cambridge, Massachusetts. David's employer pays for him to take courses at Harvard University; David has only a half-year to go toward a Bachelor's degree from Harvard's Extension School. It all adds up to savings on a valuable education that can lead in a variety of directions to a bright future. A story on David can be found in NBC2's 2011 BIOMAN News at http://www.biomanufacturing.org/bioman/BIOMAN_Journal_may201 1.pdf.

Protein is Cash High School Teachers' Biomanufacturing Workshops

The purpose of the *Protein is Cash* high school teachers' workshops is to provide hands-on activities and information about new advanced technology career paths in biomanufacturing. The hope is that these workshops will catalyze the development or expansion of the local education and training and workforce infrastructure to support biomanufacturers' need for a local advanced technology workforce. Teachers taking part in these workshops learn hands-on:

- How to transform cells with foreign genes of interest that the cell will turn into protein via the Central Dogma of Biology (DNA makes RNA makes protein);
- How these transformed cells are grown in ever increasing numbers to maximize the amount of protein of interest that is made (upstream processing);
- How the protein product of interest is purified from the cells or nutrient medium (downstream processing);
- How quality control tests are used to determine the characteristics of the protein produced (quality control biochemistry);
- How human clinical trials of the protein of interest are used to assess its safety, dosage, effectiveness, and adverse reactions;
- And how the skills learned in manufacturing a biopharmaceutical apply to the manufacture of biofuels and other products made using biomanufacturing skills and knowledge.

In addition, teachers are introduced to the design of a biomanufacturing facility, the departments within the facility, and the jobs (and career paths) associated with these areas. Teachers also hear lectures from local industry, education, and organization representatives and tour a local biomanufacturing facility.

The *Protein is Cash* high school teachers' laboratory manual, workshops, and online support site at www.biomanonline.org provide teachers with the resources needed to introduce high school students to the exciting career opportunities available in a twenty-first-century advanced-technology STEM industry, biomanufacturing.[xv] The NBC[2] is partnered with Bio-Rad (www.bio-rad.com) in offering *Protein is Cash* workshops.

Final Thoughts

Modern biotechnology has led to a new and rapidly growing STEM biomanufacturing industry. It is a young industry, just thirty years old, that will create a bioeconomy to provide us with biopharmaceuticals, replacement organs, and green products such as biofuels, bioplastics, and industrial enzymes. While the world economy still lags, biopharmaceuticals remain one of the few growth industries in which the United States (along with Europe) can claim to be the leader. In 2011, the market for biopharmaceutical proteins in the United States exceeded $100 billion for the first time.[xvi]

Over 350 biopharmaceuticals have been approved for commercial biomanufacture in the United States, and with 5600 biopharmaceuticals currently in research or process development stages, this number will grow significantly in the coming years. In addition, we have recently seen the advent and rapid growth of crossover industries such as industrial enzymes, biofuels, bioplastics, and other bioproducts.

The field of biomanufacturing remains healthy and will likely provide an increasing source of advanced technology employment.

[i] Wikipedia contributors, "Biotechnology," Wikipedia, The Free Encyclopedia, http://en.wikipedia.org/wiki/Biomanufacturing (accessed April 19, 2012).

[ii] Northeast Biomanufacturing Center and Collaborative, "NBC2 Workshops and Online Courses," http://www.biomanufacturing.org/protein_cash/manual.html (accesed April 19, 2012).

[iii] Northeast Biomanufacturing Center and Collaborative, "*Biomanufacturing Apprenticeships*," http://biomanufacturing.org/apprenticeships.html (accessed April 19, 2012).

[iv] BioPlan Associates, Inc., *Eighth Annual Survey of Biopharmaceutical Manufacturing Capacity and Production: A Study of Biotherapeutic Developers and Contract Manufacturing Organizations*, April 2011; Robert Weisman, "Mid-Size Biotechs Making Their Way," *Boston Globe*, October 23, 2011.

[v] BioPlan Associates, Inc.,*Eighth Annual Survey*.

[vi] BIO 2010, *National Biosciences Report Shows Growth*, http://www.youtube.com/v/hO8rofzHhXg (accessed April 19, 2012).

[vii] BioPlan Associates, Inc., *Eighth Annual Survey*.

[viii] Northeast Biomanufacturing Center and Collaborative, "Biopharmaceutical Manufacturing Industry Skill Standards," http://www.biomanufacturing.org/skillstandards.html (accessed April 19, 2012).

[ix] The laboratory manual is available at http://biomanufacturing.org/biocurriculum.html. Additional online materials are available at http://www.atelearning.com/BioDemo with the password "BioRad."

[x] Northeast Biomanufacturing Center and Collaborative, "Biopharmaceutical Manufacturing Skill Standards Competency Crosswalk," http://www.biomanufacturing.org/gbc2/competencies/CompetencyCrosswalk Condensed.pdf (accessed April 19, 2012).

[xi] National Center for the Biotechnology Workforce, http://www.biotechworkforce.org/ (accessed April 19, 2012).

[xii] Department of Labor, *Seventieth Anniversary Apprenticeship Video*, 2007, http://biomanufacturing.org/movies/DOLfull.wmv (accessed April 19, 2012).

[xiii] Advanced Technological Education Television, *What is Biomanufacturing?*, 2008, http://www.atetv.org/watch-videos/episode.aspx?e=1537&c=1541 (accessed April 19, 2012).

[xiv] Advanced Technological Education Television, *Community College Programs*, 2008, http://www.atetv.org/watch-videos/episode.aspx?e=1470&c=1472 (accessed April 19, 2012).

[xv] Northeast Biomanufacturing Center and Collaborative, "*Protein is Cash* Laboratory Manual," 2012, http://www.biomanufacturing.org/protein_cash/manual.html (accessed April 19, 2012).

[xvi] BioPlan Associates, Inc., *Eighth Annual Survey*.

Secondary/Postsecondary Programs of Study for Information and Communications Technicians

Gordon Snyder
National Center for Information and Communications
Telecommunications Technologies
Springfield Technical Community College, MA

From its inception, the work of the National Center for Information and Communications Telecommunications Technologies (ICT Center) has been grounded in one primary goal: to create a comprehensive and sustainable national education system for the Information and Communications Technologies (ICT) industry.

The ICT industry—driven by a demand for instantly accessible information—is profoundly transforming the world. For more than a decade, the ICT industry has been planning for the convergence of the telecommunications and information technology industries. Business and industry pundits have expected this convergence to be complex and disruptive of all aspects of the industry. The convergence is now here, and its implementation throughout ICT networks will continue to accelerate in the coming years. Changes to the technologies within the ICT industry are now both evolutionary and revolutionary, and are originating from multiple sources, not just from changes in the networks themselves. Change in current technologies is constant, and the pace of change is accelerating. Entirely new technologies are regularly being introduced and require integration into the network. Consumers are demanding instant communications across a worldwide network through an increasingly complex array of end-user devices, and this demand creates both opportunities and challenges for the industry.

Voice, data, and video communications across a worldwide network are creating opportunities that did not exist a decade ago. Preparing an appropriately skilled workforce is a major challenge for the ICT industry because of quickly changing technology. Anticipating constantly and rapidly evolving breakthroughs in technology, the ICT Center—an Advanced Technological Education Resource Center—believes that education is the key to meeting this challenge. The ICT Center's response to these high-speed developments is founded in the answers to three important questions: 1) How can ICT pedagogy—including both content and means of delivery—be kept current? 2) How can a group of highest-quality subject-matter experts be readily engaged? and 3) How can the best of this knowledge be shared and disseminated across the nation quickly?[i]

Since 1998, the ICT Center has demonstrated a creative and effective way of answering these questions and shaping ICT education. Through the continuous collection, examination, analysis, and synthesis of ICT scientific and pedagogical material and the use of innovative, up-to-the-minute dissemination methods to share best practices, the Center constantly and consistently meets ICT academic and industry needs. And what the ICT Center has learned from this intensive effort is that the universal element in the equation is people. The Center draws on the knowledge of subject-matter experts, listens to the needs identified by industry professionals, and responds to faculty requests. People are the sources, the tools, and the beneficiaries of the ICT Center's inventive program.

Industry Snapshot

According to the Department of Labor's Employment and Training Administration (DOLETA) "High Growth Industry Profile: Information Technology," the IT industry is expected to continue to grow.[ii] The profile provides a brief snapshot of the IT industry:

- "The computer systems design and related services industry is among the economy's largest sources of employment growth. Employment increased by 616,000 over the 1994–2004 period, posting a staggering 8.0 percent annual growth rate. The projected 2004–14 employment increase of 453,000 translates into 1.6 million jobs, and represents a relatively slower annual growth rate of 3.4 percent as productivity increases and offshore outsourcing take their toll.

- "However, the main growth catalyst for this industry is expected to be the persistent evolution of technology and the ability of technicians to absorb and integrate these resources to enhance their productivity and expand their market opportunities.
- "Employment of computer and information systems managers (including technicians) is expected to grow between 18 to 26 percent for all occupations through the year 2014."[iii]

Skills and training are among the workforce issues that the profile discusses: "Over 90 percent of IT workers are employed outside the IT industry, which makes it necessary for them to have complementary training in their respective business sectors such as health care, manufacturing or financial services. Employers are also looking for well developed soft skills, transferable IT skills and adaptability in their workforce. Incumbent training programs may help in this respect, as could community colleges."

The profile draws on the U.S. Bureau of Labor Statistics 2006-07 Career Guide to Industries to broadly describe skill sets:

- "For all IT-related occupations, technical and professional certifications are growing more popular and increasingly important.
- "IT workers must continually update and acquire new skills to remain qualified in this dynamic field. Completion of vocational training also is an asset. According to a May 2000 report by the Urban Institute, community colleges play a critical role in preparing new workers and in retraining both veteran workers and workers from other fields.
- "People interested in becoming computer support specialists typically need an Associate degree in a computer-related field, as well as significant hands-on experience with computers. They also must possess strong problem-solving and analytical skills as well as excellent communication skills because troubleshooting and helping others are such vital aspects of the job. And because there is constant interaction on the job with other computer personnel, customers, and employees,

computer support specialists must be able to communicate effectively on paper, using e-mail, and in person. They also must possess strong writing skills when preparing manuals for employees and customers."

Clearly, this diverse industry would be well served by a national community of thought leaders who represent as many aspects as possible of this fast-changing and evolving industry.

Partnerships

The ICT Center has become a critical resource for faculty and administrators working to create, modify, and strengthen ICT programs. Requests, often in response to the rapid pace of technology, have included assistance with developing content and associated labs, identifying lab equipment, designing appropriate curricula, acquiring and using industry input, and identifying and assisting with the development of institutional program funding. The Center has provided staff knowledge and experience, and by offering invitations to attend ICT Center conferences, has also provided access to faculty and administrators and their expertise.

ICT Center staff gained two key insights from these interactions and from annual conferences. The first was that the core staff could not adequately respond to the volume of requests for assistance, nor could they gain and maintain the expertise needed to keep pace with the rapid changes in technology, the complexity of these systems, and the diverse regional needs. Second, it became very clear that the Center's annual conferences, which provide educators and attendees access to peers and an opportunity to network with their colleagues, have become an invaluable resource to participants.

As a result, the Center gathered an informal national network of ICT educators and business representatives into a Regional Partner Network, a group of thought leaders from academia and business who share ideas and experience and together develop solutions to problems and issues in their respective endeavors. These partnerships are a key component of the Center.

Curriculum Evolution

The recovery of the information technologies (IT) and telecommunications industries from the dot-com bust of 2001 and 2002 resulted in the new and emerging technologies that are enjoyed today. These changes occurred as companies—the telecommunication giants as well as the IT start-ups—merged or reorganized to create a new converged industry. In 2003, the Center, academia, and most of business and industry were still focused on traditional telecommunications, including copper, wireless, and optical-based delivery systems and networks.

Through its work with the Verizon NextStep program and other industry partners, the Center predicted the convergence of voice, video, and data services over Internet-technology-based networks, and responded early and rapidly to broaden program coverage to include these emerging technologies. The ICT Center expanded content coverage to include end-device–to–end-device communications and trademarked the term "Connecting Technologies" to better describe these rapidly emerging networks. To better address this expanded focus, new subject-matter experts were brought in to help with this new content. The ICT Center Regional Partner Network played a critical role in helping the Center rapidly acquire and modify content and subject matter expertise from academic and business partners.

Targeted Academic Workshops and Conferences

The ICT Center workshops began as a gathering of teachers and faculty who came to learn content relevant to telecommunications technician education. The Center's conferences now reflect not only technology change, but also the needs of the different regions of the country. For example, the conferences now address content development in response to the needs of local companies, program-improvement resources, and networking and sharing ideas among colleagues. A grant-writing session was added to explain funding opportunities and effective responses to grantor proposal requirements.

The Center circulated a call for presentations in new content areas to its national LISTSERV of academic and business professionals. These presentations give conference attendees materials for immediate use in their classrooms that are available on the Center website and YouTube channel. The ICT Center has recognized that IT and telecommunications are merging to become ICT technologies. As a result, faculty, business and industry professionals, and employees have recognized the ICT Center conferences as a forum for the exchange of ideas, the sharing of experience, and the development of solutions for the faculty and administrators who attend from around the country.

Business and Industry Relations

The ICT Center has engaged key ICT business and industry organizations in a range of activities, including conferences and seminars for professionals within the field, curriculum development, advising, student internships and externships, and Center conferences for educators. Today, the ICT Center has active relationships with over seventy corporations. This type of sharing of information and concepts supports the ATE program goal of enabling best practices in technical education to become standard operating procedures for educators throughout the United States.

The Center has been closely involved in the Verizon NextStep program since NextStep's inception. The NextStep Program allows qualified Verizon associates who are members of the Communications Workers of America or the International Brotherhood of Electrical Workers to earn an Associate in Applied Science degree in Telecommunications Technology from a partnering college. Since starting in 1996, this program has had over seven thousand Verizon employee participants.

The Center has also established a relationship with the North Central Division of Comcast, and is currently engaged in the development and piloting of curriculum and training materials for the two new positions at the top of their technician career ladder. These new positions were created specifically to support the company's transition to its new Internet-protocol-based (voice, video, and data) network and services. The North Central Division may serve as a pilot for eventual national use.

Authoring Content

It has been difficult for textbook publishers to keep traditional textbook content up to date; as a result, the Center expanded its content dissemination. In 2005, the Center began using Web-2.0-based technologies, including blogs, audio podcasts, video podcasts, a YouTube channel, and a textbook wiki. These tools have proven to be an effective way for faculty and students to keep up to date with current and emerging technologies. The ICT Center uses interactive dissemination tools to reach over one million users annually:

Twitter, Facebook, and Google+: The Center's Klout.com impact score translates to 1.8 million contacts of ICT-related content per year.

YouTube: By September 2011, the ICT Center's videos had been viewed 77,301 times; its channel had been viewed 9,269 times and had 120 subscribers.

Podcasts: By September 2011, the Center had created and distributed 105 audio and video lectures; they had been downloaded 120,991 times.

Blogs: ICT Center online blog content is viewed 138,000 times every year.

The ICT Center Community of Practice

In 2008, the Center focused its efforts on formalizing the Community of Practice (CoP), which brings together an expanding group of academic professionals and industry experts who share a common goal of ensuring a high-quality and industry-relevant education for all ICT students. The CoP was designed to continue to grow and mature as organizations change and new technologies are developed.

Through its ICT Community of Practice, the Center continues to build organizational capacity, expand regional and demographic representation, and broaden subject-matter expertise to ensure further cross pollination for content development. The Community of Practice acts as a a flexible system that makes it possible to identify business and industry workforce needs while communicating those needs to curriculum developers. Finally, the Community of Practice helps the

Center maximize the dissemination of all its collaboratively developed materials and expertise. These activities have strengthened the Center over the past three years, attracting and supporting future involvement by motivated faculty, institutions, and business and industry professionals.

The modern ICT-technician workforce requires knowledge of both the hardware and the software involved in end-device–to–end-device communications, including the specifics of delivery systems and customer-site equipment.[iv] Today's ICT technician needs to know more than just how to apply skills to a particular task. Today, where, why, and when are also important to systems design and troubleshooting. Core knowledge of technology principles and the application of skills formed in a business and industry context are particularly crucial to the success of ICT technicians as the ICT industry continues to integrate new and emerging technologies into its operations.

Many ICT-technician programs have not included the math and science foundation required of ICT technicians working in the new, converged space. Further, many technicians find themselves dead-ended in their careers and may need to begin an alternative career path that is aligned with the technology changes affecting the industry. Professional development programs for ICT faculty must keep pace with industry changes and establish a process to adapt their curriculum and training programs. Institutions that regularly incorporate these changes into their technology programs stay current with industry requirements and avoid program obsolescence.

Evolution of the ICT Center's Approach

Initially, the ICT Center was established as a regional center to create a network of academic institutions focusing on traditional telecommunications, including wireless, light-wave, and networking. Mergers and acquisitions within the industry combined legacy technologies with new technologies. Rapid technological change and conversion to Internet-based technologies for delivery of voice, video, and data communications have had a profound and disruptive impact on business and industry. In the early 2000s, this dynamically changing landscape required the Center to expand and develop new goals.

In 1991, Etienne Wenger and Jean Lave first used the term "community of practice" to signify a process of knowledge management.[v] Lave and Wenger refer to "knowledge being the key source in the business world, with the limitations of most having little understanding of how to create and leverage it in practice."[vi] Lave and Wenger go on to state that "systematically addressing the kind of dynamic 'knowing' that makes a difference in practice requires the participation of people who are fully engaged in the process of creating, refining, communicating, and using knowledge."[vii]

This process of creation, refinement, rapid and broad dissemination, and use of knowledge symbolizes the evolution of the Center from what it was in 2002—three subject-matter experts located on the STCC campus—to its current incarnation as a group of hundreds of subject-matter experts (SMEs) scattered across the United States currently in what Lave and Wenger refer to as the Coalescing Phase.[viii]

With a large collection of eager, competent SMEs, and business and industry partners who have established trusting relationships, this group is poised to move to the next phase in the Lave-and-Wenger cycle: the Active Phase.[ix] Members of this emerging community have built strong relationships—including new industry relationships with, for example, Comcast, Juniper, and Nortel—and are growing their own organizations. Three members of the community have previously been funded by the NSF, and four are currently funded by the NSF. Two members submitted funding proposals to the NSF in October 2011.

Lave and Wenger define a community of practice along three dimensions:

1. What is a community of practice about? "A community of practice is a joint enterprise as understood and continually renegotiated by its members."[x] The Center has continually refined its process to better reach individual faculty members who have the desire and ideas to make academic programs better; the Center has then helped these faculty members share their successful changes with others.

2. How does a community of practice function? "A community of practice functions through mutual engagement that binds members together in a social entity."[xi] The Center has always realized the power of its network and the importance of connecting members together in ways that provide two-way support, expertise, and advice. By providing access to those who have "been there before" and succeeded, the ICT Center makes the appropriate connections and accelerates the development process.

3. What capacity does a community of practice produce? "A community of practice produces a shared repertoire of communal resources that includes the routines, sensibilities, artifacts, vocabularies, and types that members have developed over time."[xii] The Center's resource library includes voice, video, and data content and was formed using the content-sharing model created by Creative Commons, a copyright rights nonprofit. This library enables Center affiliates to rapidly find what they are looking for and connect with the developer of the content.

Early-stage communities of practice in the ICT Center organization, although not formally referred to as such, have been foundational in the Center's growth and success to date. Because membership is based on participation, the community is not bound by organizational affiliations and can span institutional structures and hierarchies.

Structure of the Community of Practice
The members of the CoP form its core and are the center of its expertise. Knowledge is developed at the core, and each member or node facilitates the exchange and local or regional interpretation of information. The CoP model is an ideal mechanism for moving information; however, it is difficult for the core to develop new expertise. External expertise and stimuli provide this input of new knowledge through interactions at the boundaries of the CoP. A permeable boundary is critical to the success and continued renewal of the CoP because it creates opportunities for learning and new insights by involving newcomers and inviting external expertise.

To work effectively, communities of practice must be self-organizing, and participation must be intrinsically self-motivated and self-sustaining. The shared learning and interest of its members are what maintain the CoP. In fact, the CoP exists because the members derive value from participation in a dynamic, forward-looking community. The organization can support participation by providing access to outside experts, as well as resources for travel, meeting, and communications.[xiii]

Often, faculty are working either unrecognized or invisible at their home institution or are doing work that is only recognized by a small group and not officially sanction by the institution—Lave and Wenger refer to this as a "bootlegged relationship."[xiv] Participation in the CoP has additional benefits at the partners' home institutions: it helps legitimize their work, provides strategic benefit that may lead to increased resources, and it helps them modify and update their curriculum and programs.

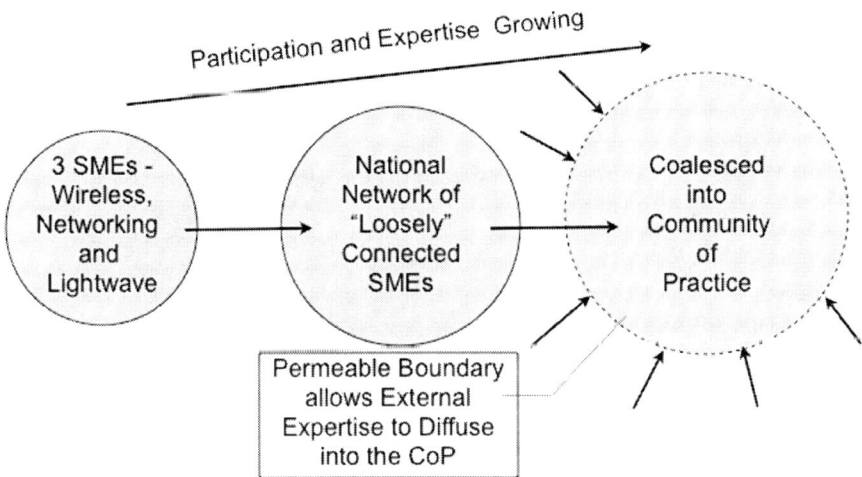

Figure 8.1. Community of Practice Structure

The formalized ICT-Center CoP has served as a comprehensive structure that connects faculty with similar needs and varying levels of expertise using the community-college mission of responding to regional workforce needs. The Center has become a catalyst for a community of practice linking educational institutions, business organizations, and the ICT workforce to develop consensus on best ICT content. Additionally, this community of practice will position the Center to stay abreast of current and future trends, and will be a market watch for existing and emerging technologies. According to Lave and Wenger, the successful implementation of a CoP will result in:[xv]

- Students getting the core knowledge of technology principles that will better prepare them to work in the changing ICT industry.
- Technicians having a better understanding of the skills and knowledge they need to gain or advance in employment, making more informed decisions about their career path and skill development opportunities, and being more proactive in seeking relevant career development information.
- Industry recognizing the Center as a long-term, strategic education partner whose primary focus is the better alignment of workforce and workforce training activities to meet the demands of their future technology innovations.
- Educational Institutions and interested faculty knowing that the Center is the most cost effective and reliable source for information and support to start, upgrade, or refresh their ICT programs.
- The ICT Center supporting an increasing number of participating educational institutions and businesses without commensurate growth in the organization; maintaining relevancy of current content and materials to industry needs; responding to new technologies and technology elements within its focus in a timely manner; and improving the pace and breadth of dissemination of its products.

The CoP continues to grow as others sharing similar interests learn of the community through Center conferences, related industry and academic conferences, expanding business relations, and the ICT Center's online presence. According to Lave and Wenger, some of the short-term and long-term values of a CoP include: keeping abreast of developments in the field, fostering innovation, solving problems, promoting organizational intelligence, building connections, retaining talent, and developing new strategies. These and other characteristics of a CoP allow new members to quickly gain access to the CoP's collective knowledge and bring that information to their own institutions.[xvi]

A brief story highlights the success of a potential CoP by demonstrating what the Center's academic partnerships have spawned. Under the guidance of the ICT Center and through an NSF grant, two California regional partners came together to upgrade courses and develop a new degree program at one of the institutions. The Institute for Telecommunications Technologies at Cuyamaca College provided support, advice, and materials to the City College of San Francisco (CCSF) for the formation of the Institute for Convergence of Optical and Network Systems (ICONS). The NSF project grant was a result of the principal investigator having attended ICT Center Regional Partner meetings and a summer conference that provided grant-writing information and training, as well as having formed connections with other regional partners. Additionally, ICONS created programs to improve outreach to historically underrepresented groups of students in the San Francisco area. Eventually, ICONS transitioned into a regional center serving the mid-Pacific region; it is now the Mid-Pacific Information and Communication Technology Center.

Modern ICT Community of Practice

Today, many community-college ICT and ICT-enabled programs are challenged by the need to make extensive and fundamental curriculum changes with little support or direction. The ICT Center Community of Practice provides program assistance, a forum for the dynamic exchange of ideas, methods for interactions, and a dynamic resource set where current and new content, subject-matter experts, and support are readily available and accessible. The ICT Center has created a Community of Practice (CoP) that:

- provides readily accessible and relevant content to academic and business professionals nationally,
- ensures that students receive and use classroom-ready materials that reflect current and emerging technologies,
- directs members' exchange of ideas and experience into focused activities,
- increases the number of ICT subject-matter experts,

- provides focused business and industry participation involving regional and national companies,
- reaches out to underrepresented groups to encourage their participation in the field,
- continues to expand the capacity of the organization, and
- provides resource support to CoP members and helps identify funding sources for program improvement and development.

The Center continues to build on its program success by solidifying, formalizing, and expanding its Community of Practice. The ICT Center believes that a strong ICT Community of Practice assures the crux of its goal—a comprehensive and sustainable ICT educational system.

Preparing ICT Technician Students at STEM High Schools

The figure below shows a 4+2 (secondary/postsecondary) curriculum structure that will serve modified STEM high schools working with AAS degree ICT programs at community and technical colleges.

Soph 2		Elective	Humanities		Computer Hardware and Operating Systems 2	Cisco Networking 4	Computer Networking 2
Soph 1	Elective		Social Science		Computer Hardware and Operating Systems 1	Cisco Networking 3	Computer Networking 1
Fresh 2	College Algebra			Physical Science	Computer Applications 2	Cisco Networking 2	Intro to Computer Programming 2
Fresh 1		College English		Physical Science	Computer Applications 1	Cisco Networking 1	Intro to Computer Programming 1
12th Grade	Algebra 2 w/Trig	English 12	Government	Physics	Health	Introduction to TCP/IP	Unified Networks (Data, Voice and Video)
11th Grade	Math Applications	English 11	American History	Chemistry	Physical Education	Communication Systems	DC-AC Electricity
10th Grade	Geometry	English 10	World History	Biology	Physical Education	Foreign Language	Principles of Engineering
9th Grade	Algebra 1	English 9	Geography	General Science	Physical Education	Foreign Language	Intro to Engr Design

Figure 8.2. 4+2 Curriculum Sequence for ICT

This sample curriculum can be augmented and customized depending on the direction the student wants to pursue. For example, a student that is interested in taking a software development route would want to take additional courses in topics including C, Java, and mobile application development. A student that is more infrastructure-focused would want to take additional courses to concentrate on voice, video, and storage.

The course titles used in this curriculum were chosen to represent topics that appear in typical ICT AAS programs built from an electronic core curriculum. This is a "model curriculum"—a starting point from which high schools and two-year colleges can begin working together to build an ICT career pathway for their students that meets state educational requirements and satisfies the skill standards required by local employers.

As pointed out in chapter 4, one way of building a robust student pipeline through a career pathway is to provide within the curriculum opportunities for students to take dual-credit courses. The ICT model curriculum offers this opportunity. The four Cisco specialty courses are standard in AAS networking curricula and would earn students college credit at most community colleges upon successful completion if taken at the high school level. Depending on the capability of the student, there is also a potential within the model ICT curriculum for students to earn dual credit for mathematics, science, and English courses as well. The financial and motivational implications of accelerating a student's education in this way are obvious.

Though building a standards-based, state-approved curriculum with dual-credit opportunities is a viable means for attracting students to an ICT technician career pathway, the curriculum, by itself, is not enough to ensure that the student pipeline will stay full and flowing. "Pipeline building" must also include making students aware of career opportunities in ICT and motivating them to seek ways to become a part of this field.

As chapter 4 indicated, generating awareness about technical careers can begin as early as the elementary grades. For example, many elementary students have been exposed to mobile technology, and many are using mobile devices such as smart phones and tablets on a daily basis for voice, video, and data services. For many students, there is already an awareness and interest in ICT. Effective pipeline building provides students a means to nurture this interest and determine if information technology would be a good career choice for them.

---

[i] Mike Qaissaunee and Gordon Snyder, "The Future of Mobile Teaching and Learning"(presentation transcript, April 2009) http://www.slideshare.net/mqaissaunee/the-future-of-mobile-teaching-learning (accessed May 2012).

ii United States Department of Labor, "High Growth Industry Profile: Information Technology, http://63.88.32.17/brg/Indprof/IT_profile.cfm (accessed May 9, 2012).

iii Ibid. All quotes in this section come from this Department of Labor Web page.

iv Dave Wilcox, "Rebuilding the IT Organization for Convergence," *Business Communications Review*, July 2007

v Jean Lave and Etienne Wenger, Situated Learning Legitimate Peripheral Participation (New York: Cambridge University Press, 1991).

vi Ibid., 1.

vii Ibid.

viii Ibid.

ix Ibid.

x Ibid.

xi Ibid.

xii Ibid.

xiii Ibid.

xiv Ibid.

xv Ibid, 1-2.

xvi Ibid.

Secondary/Postsecondary Programs of Study for Biotechnology Technicians

Elaine Johnson
Bio-Link Next Generation National
Center for Biotechnology and Life Sciences

What is Biotechnology?

People have used living organisms to produce commercial products for hundreds of years. But biotechnology as we talk about it today refers to the commercialization of the recombinant DNA techniques discovered by Herbert Boyer and Stanley Cohen in 1973. Recombinant DNA is created by joining genetic material from two different sources. This usually involves taking a gene from one organism and enzymatically cutting it out and inserting it into the DNA of a different organism. The beauty of this technology is that it allows transfer of DNA across species. An example of this technique is the insertion of the gene for human insulin into bacterial DNA. The bacteria act as miniature factories and make it possible to produce large quantities of human insulin.

We use restriction enzymes to cut the double-stranded DNA, and we purposefully cut "unevenly"—that is, we cut the two strands at difference places. This process creates "sticky ends" on the cut fragments. When these join with host DNA at a site that has also been enzymatically cut, the sticky ends are able to join and be incorporated into the host DNA. This technology permits the relatively inexpensive production of products of commercial value. The techniques require skilled technicians who work under regulated conditions. These technicians need to understand aseptic technique, cell growth conditions, measurement of cell concentrations, and production output.

Technicians need to document their work by recording procedures and results using standard practices.

It took almost a decade to bring this new technology to market with the production of the first clinically approved therapeutic human insulin in 1982. Since then, there has been an explosion of new techniques and applications that have revolutionized the world of possibilities for growing the bioeconomy and providing opportunities for new technical career pathways. It is the emerging biotechnologies that are driving the workforce development need for programs to prepare technicians with the skills required by the growing biotechnology industry.

Historically, biotechnology can be divided into the period prior to 1981 and the period following 1981, which featured the rapid development of new technologies, applications, and commercial use of recombinant DNA technology. Prior to 1981, applications of biological activities to commercial production of usable products included the making of wine, beer, cheese, yogurt, and many other commodities. The story of the production of penicillin from mold is now well known. But these products are not genetically modified and are not part of the "new" world of biotechnology. Since 1981, the biotechnology industry has been using tools and techniques developed to utilize recombinant DNA technology for the manufacture of useful products and applications. We are now experiencing the growth of companies that are producing medicines, biofuels, gene therapies, stem-cell applications, forensics, pest-resistant plants, improved nutritional products, vaccines, biosecurity assays, environmental testing, personalized medicine, synthetic biology, and countless other new applications. Because most of these are highly regulated, in addition to technicians who actually do the research and production, there is a growing requirement for regulatory and quality technicians. Increasingly, community and technical colleges throughout the nation are implementing programs to provide relevant competency-based education that will prepare these biotechnology technicians.

Since the birth of the biotechnology industry at Genentech in 1976, the number of biotechnology companies has exploded. In the San Francisco Bay area alone, there are over nine hundred life-science companies; together, these companies are responsible for ninety thousand jobs that require a bioscience background.[i] From 2003 to

2008, employment in the life sciences has climbed steadily in California at a rate of 3.2 percent in Northern California and 2.3 percent in Southern California.[ii] Bioscience is one of the fastest-growing industries in the nation, with a high demand for entry-level workers holding two- and four-year college degrees. The bioscience employment base reached 1.3 million jobs in the United States in 2006 with the addition of nearly fifty-three thousand jobs between 2004 and 2006, and every bioscience job creates 6.2 additional jobs.[iii] According to a report released by the research organization Battelle, the biotech industry is "led by strong growth in the research, testing and medical lab subsector, which experienced a 17.8 percent increase in employment and a 32.7 percent increase in establishments between 2001 and 2006." Looking toward the future, a 2009 Burrill report entitled "Life Sciences: A 20/20 Vision to 2020" reinforces the immense potential and rapid growth of the biotech industry.[iv]

At the same time, the bioscience industry is rapidly changing to include working relationships with other scientific disciplines and technologies. In the words of an earlier Battelle report, "Just as information technology drove economic progress in the latter part of the 20th Century, the convergence of advanced technologies in information technology, engineering, and biological sciences is producing widespread [new] opportunities."[v] It is clear that workforce training will become more interdisciplinary as convergence of disciplines continues to take place.[vi] Furthermore, concentrations of bioscience sectors within regional economies are now beginning to drive the development of focused technical-education programs at community and technical colleges. According to the U.S. Bureau of Labor Statistics, job opportunities for biological technicians are expected to grow by 28.2 percent between 2004 and 2014, while the those for biological scientists is projected to grow by 17.0 percent.[vii] In light of the extensive growth and dynamic, interdisciplinary nature of the biosciences industry, community and technical colleges face the significant challenge of preparing large numbers of students for employment while keeping abreast of rapidly changing workplace opportunities and requirements. The state of North Carolina, as one single example, addresses growing biotechnology statewide; in North Carolina, biotechnology represents a $64 billion economy and 226,000 total jobs.[viii]

In spite of the increased demand for skilled technicians, students still find it difficult to make wise decisions about their careers and often do not understand the relationship between their skills and the full range of technical employment opportunities. The primary reasons for this predicament are: (1) research institutions and industries employ multidisciplinary problem-solving approaches with technology playing a key role, but multidisciplinary, technology-rich programs are not yet emphasized at many community and technical colleges; (2) the connections between educational programs, the biosciences industry, and related career options are not well understood by many students; and (3) community and technical college biotechnology programs are typically conducted in relative isolation, without a robust national system for sharing information and intellectual resources (a situation that Bio-Link, a center created to enhance education in biotechnology, is helping to change).

Bio-Link: Advanced Technological Education Center of Excellence

The Bio-Link Next Generation National ATE Center for Biotechnology and Life Sciences builds on the success of the original Bio-Link National Center for Biotechnology that was first funded by NSF in 1998. Bio-Link's mission is to:

1. increase the number and diversity of well-trained technicians in the workforce;
2. meet the growing needs of industry for appropriately trained technicians; and
3. institutionalize community college educational practices that make high-quality education and training in the concepts, tool, skills, processes, regulatory structure, and ethics of biotechnology available to all students.

Bio-Link has made great strides in developing biotechnology faculty, disseminating curricula and laboratory resources, providing accessible information about educational and employment opportunities, and helping establish formal and informal biotechnology faculty networks. However, with the rapidly accelerating growth of the biotechnology industry and the convergence of the biosciences with other disciplines, technical applications and job requirements are changing swiftly, and Bio-Link must adjust its goals and activities to better prepare students for this reality.[ix]

Community and Technical Programs for Biotechnicians

In the late eighties and early nineties, biotechnology programs began to emerge in different areas of the country, with Madison Area Technical College in Wisconsin and Alamance Community College in North Carolina leading the way. New England, California, and Washington followed close behind. At about the same time, high school teachers began including some basic biotechnology experiments in their science classes. Bio-Link began collecting information about these programs in 1999. At that time, it was not evident which two-year colleges had biotech programs. Biotechnology faculty members were working in isolation and were independently developing programs to meet local workforce needs. Bio-Link surveys were sent to every community and technical college in the nation. In 1998-99, forty-nine formal programs were identified, and by 2005-06, the number had grown to ninety—an 84 percent increase with 57 percent growth between 1998-99 and 1999-2000.[v] Our current preliminary data show eighty-six colleges with biotechnology programs.

Bio-Link has spent the last fourteen years connecting the biotechnology technical workforce development community. During that time, Bio-Link transitioned to a resource center and then back to a National ATE Center of Excellence. All the while, Bio-Link has systematically conducted surveys to identify characteristics of those community and technical college biotechnology programs that are partnering with local industry to prepare the technical workforce for the biotechnology industry. The results of the surveys over the years have provided rich data to support the critical role of community and technical colleges in technician education. Bio-Link is currently gathering another set of data, and preliminary indicators show a slight increase in the number of programs, with several new ones in the Southeast region of the nation. As a result of the Bio-Link network, community and technical colleges across the country have a community of practice to support the continued quality and currency of their programs.

Many states have developed their own networks. In North Carolina, BioNetwork combines campuses across the state into a life-science training enterprise that also includes seven specialized centers. Job seekers and new and incumbent workers at all levels can learn new skills through short courses, certificate programs, and associate degree programs.[xi] Similarly, the Seattle metropolitan area is one of the nation's leading employers of laboratory specialists and features over one hundred biotechnology-related facilities. In 2000, Shoreline Community College created a set of Biotechnology/Biomedical Skill Standards that have influenced biotechnology curriculum and articulation at high schools and community colleges in the northwest and other regions of the country, and they updated these standards in 2007.[xii] Several Biotechnology Skill Standards can be found at the Bio-Link Curriculum and Instructional Materials Clearinghouse at http://www.bio-link.org. Additional examples include the work of BioOhio, MassBio, BayBio, BioCom, GeorgiaBio, MDBio, and other local trade organizations that are actively engaging with their local community colleges to improve workforce education for the life-science industries. The BayBio *Impact 2010* report recommends sustaining funding for California community colleges to recognize their role in shaping education and preparing the technical workforce of the future.[xiii] Many more examples could be provided.

U.S. community and technical colleges have a staggering amount of human capital with an enrollment of more than 12.4 million students (44 percent of the U.S. undergraduate student body).[xiv] Over the last decade, Bio-Link has recognized the tremendous value of utilizing this pool of students for engaging and preparing the skilled bioscience technical workforce.[xv] Community and technical colleges are positioned to address the changing needs of the workforce and are convenient for students who wish to take advantage of their offerings. A strong case can be made for aligning higher education with workforce needs.[xvi]

Industry Needs for Five Years

As more and more programs developed, it became clear that there were certain barriers that kept people from choosing pathways that led to technician careers in biotechnology. A conference sponsored by the

National Science Foundation (NSF) and the American Association of Community Colleges (AACC) was held in Scottsdale, Arizona, April 28–30, 2008, for the purpose of addressing industry needs for the next five years.[xvii] A call to eliminate barriers was a strong message that permeated the meeting's lively discussion. The consensus of the fifty attendees was that every specialization and emerging area needs technicians with strong fundamental technical skills who are also able to be flexible and adapt to changes in the workplace.

Recommendations for biotechnology curricula included soft skills, core courses that transfer, strong theoretical understanding of the manufacturing process, the introduction of emerging technologies in basic courses, and the redesign of standard curricula to include applications in biotechnology. Since this meeting, there has been a major downturn in the economy, but the biotechnology industry is weathering the storm. In 2011, *Shaping Infinity: The Georgia Life Sciences Industry Analysis* described life-science industry employment as "recession resistant." The report singled out the need to find and hire skilled technicians as one of the most important labor force factors.[xviii] Another report, entitled *Taking the Pulse of Bioscience Education in America: A State-by-State Analysis,* supported this conclusion with the statement that "the bioscience industry is a knowledge-based sector dependent upon the skills of its workers. Thus, ensuring the availability of an educated, skilled workforce is critical to developing and sustaining a highly competitive, robust bioscience cluster over the long term."[xix]

It is not surprising that the federal government is interested in biotechnology and the role of community colleges in preparing skilled workers. The Science and Technology Policy Office of the White House has asked for public input into decision making to support the bioeconomy. A National Bioeconomy Blueprint will detail steps to use biological innovations to address national challenges in health, food, energy, and the environment. The White House is requesting suggestions about what kinds of investment in education and training are essential to creating a technically skilled twenty-first-century American bioeconomy workforce, and is specifically requesting input about the roles of community colleges in training the bioeconomy workforce of the future.[xx]

Biotechnology programs that prepare the technically skilled workforce for the biotechnology and related life-science industries fit hand-in-glove with the growth of the bioeconomy. With the ongoing support of the NSF/ATE, Bio-Link continues to provide a connecting point for biotechnology education across the nation. There is a critical need to sustain these programs and take advantage of the innovations and collaborations that have been established. It is more critical than ever that information about careers and how to gain skills valued by the industry be made available to students, teachers, counselors, parents, industry partners, and all stakeholders.

Career Pathways for Biotechnicians

Students can be introduced to biotechnology careers at an early age. By middle school and high school, students are making choices about the direction of their education. We must set the goal of keeping students of this age interested in life science. The new Georgia high school biotechnology curriculum has been succeeding at this goal: students report "positive changes in key benchmarks, including: students' interest in continuing science courses; confidence in their ability to tackle more advanced content; motivation to be successful in class, solve problems, and find answers to challenges; and expectations that biotechnology knowledge will benefit them in their future careers."[xxi] Many regions of the country are including biotechnology laboratories in science classes, and many have entire courses devoted to biotechnology. One common element that makes such courses successful is support for teachers through professional development, materials, supplies, and troubleshooting.[xxii] There are some stellar examples of remarkable teachers who write materials, run workshops, teach their course as if it were a business, give students forensics or research exercises, and generally engage students in the excitement and potential applications of biotechnology. Outstanding resources for high school teachers can be found at http://www.bioteched.com.

Articulation of biotechnology high school courses to community and technical college biotechnology programs is scattered. Salt Lake Community College provides one example of strong articulation and dual enrollment for high school students. City College of San

Francisco and Austin Community College provide different models. Many high schools and community colleges around the nation are also creating articulation agreements. Unfortunately, articulation is not the only issue. It is estimated that only half the nation's high school graduates are prepared for postsecondary academic work. As in the report *Thriving in Challenging Times* puts it, "Yesterday's educational system is inadequate to meet today's realities."[xxiii] Many students are simply not equipped for future success in STEM careers. Without the necessary tools, they are not able to move confidently into biotechnology programs. The Bridge to Biotech process that I describe later in this chapter offers an alternative to traditional remedial courses.

Surprisingly, large numbers of students with baccalaureate degrees come to community and technical colleges to receive the hands-on skill training required to enter the workplace. This is true for recent graduates with molecular biology degrees from quality institutions as well as for baccalaureate degree holders with non-science degrees who want to enter the exciting field of biotechnology.[xxiv] The following example illustrates that the pathway is not always straight and that there are multiple entry points into programs that provide outstanding biotechnician education.

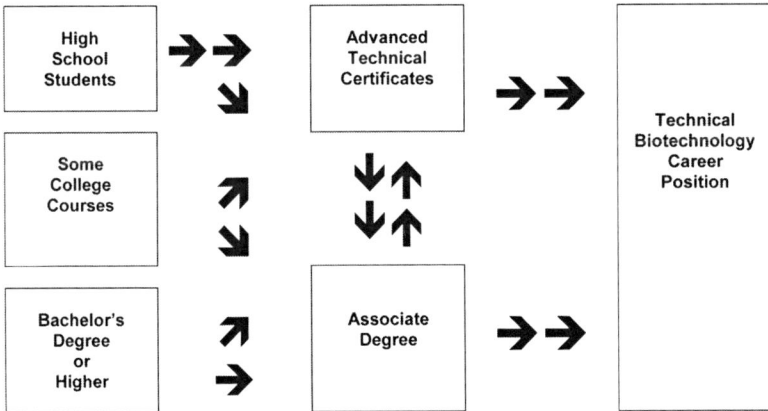

Figure 9.1. Multiple pathways leading to biotechnology programs and placement.

Some, but not all colleges, offer an Associate of Applied Science. Many offer an Associate of Science or an Associate of Arts. This varies from state to state and even from college to college within some states. This variation is one of the reasons that industry-endorsed portable skills certificates are being considered. In this way, as graduates from the community or technical college biotechnology programs move to a different location, they carry with them a standard, industry-recognized, portable certificate.

The entry point into community and technical programs is varied. The diversity of students extends beyond ethnicity and age. The backgrounds of people are also very different. High school students can benefit greatly from entering programs with mature people who can share their life experiences as well as be classmates learning skills and science. This is a value added for many students who are entering biotechnology programs directly from high school.

Soph 2	Pre-calculus or Elective	Technical Writing	Humanities Bioethics	Technical Biotech Elective	Internship	Biotechniques Specialties	Fermentation & Protein Purification
Soph 1	Statistics		Social Science	Conceptual Physics		Biotechniques Specialties	Molecular & Cell Biotechnology
Fresh 2	College Algebra	Speech		Chemistry	Intro to Bioinformatics	Biotechniques Specialties	Recombinant DNA Biotechnology
Fresh 1	Contextual Math	College English		Chemistry	Biology/ Health	Research Skills	Biotech Industry Lecture Series
12th Grade	Algebra 2 w/Trig	English 12	Government /Economics	Physics	Health	Careers in Biotech	Internship
11th Grade	Math Applications	English 11	American History	Chemistry	Physical Education	Forensics	Biotechnology Careers and Techniques
10th Grade	Geometry	English 10	World History	Biology	Physical Education	Foreign Language	Biotechnology Techniques and Applications
9th Grade	Algebra 1	English 9	Geography	General Science	Physical Education	Foreign Language	Intro to Biotechnology

Figure 9.2. A Secondary/Postsecondary Career Pathway for Biotechnology

Data- and Industry-Driven Curriculum

New technologies and the intersection or merging of disciplines create challenges for educators and industry alike. Advanced technologies and scientific disciplines are increasingly converging; at the same time, online communities are quickly gaining a foothold as a means for communication and cyberlearning. Bio-Link has been able to create networks that connect individuals and institutions across the nation. These networking activities remain essential; however, biotechnology, other life sciences, and related fields are now changing so quickly that the personal networking model is no longer fully adequate as the primary means for fulfilling Bio-Link's mission of strengthening and expanding the biotechnology technician workforce.

The Bio-Link Instructional Material and Curriculum Clearinghouse

Biotechnology instructors at community and technical colleges come from a wide variety of backgrounds. Although some have worked in industry settings, many have only worked in academic labs. The lack of industry experience can be problematic when these faculty members are called upon to design courses that are intended to prepare students for the nonacademic workforce.

Clearinghouse materials allow instructors to access others' expertise when it comes to subjects that aren't taught in most college programs, such as good manufacturing practices (GMP), FDA Regulations, and bioprocessing. In progress are the creation of "courses in a box" that provide comprehensive materials for entire course offerings. The clearinghouse also provides references to published materials. For example, there are references to several publications by Lisa Seidman and colleagues; these publications offer basics suitable for community college biotechnology courses.[xxv]

The clearinghouse staff encourages contributions, especially from NSF-ATE funded projects. They also request feedback to assist in continually improving this resource and meeting the needs of the users.

Bio-Link Equipment Depot

To meet the growing need for science laboratory experiences, the biotechnology industry, together with Bio-Link, has established a highly successful distribution center for donated supplies and equipment. We have a dedicated site provided by City College of San Francisco, an oversight management by Bio-Link, well in excess of thirty regular donor companies, close to two hundred regular teachers from Northern California, and an experienced coordination team with student interns. We also have volunteers from BayBio Institute, the Bay Area Biotechnology Education Consortium (BABEC), interested business people, and many teachers.[xxvi] The depot has served over eighty-five thousand students. This distribution center has become more than a place for receiving expensive supplies and equipment that are beyond the reach of school and college budgets; it also serves as a gathering place for sharing information and collaborating on projects (some of which have additional funding). The depot also serves as a nexus for interaction between educators and industry.

Bio-Link Strategies

As the Bio-Link Next Generation ATE Center for Biotechnology and Life Sciences continues to work closely with community college biotech programs and industry on issues of sustainability and institutionalization, Bio-Link is addressing:

- Strategies for establishing broad company buy-in to partnerships with community college biotech programs, so that collaborative efforts do not suffer when there is turnover in company or college personnel.
- Ways in which biotech faculty can coordinate and manage programs that include a strong component of industry participation.
- Approaches to ensuring the support of community college administrators for in-depth cooperative relationships between their college biotech programs and local life-science companies, particularly when the relatively high cost of biotech courses may discourage budget-limited colleges from offering a wide range of selection of courses.

- Opportunities for college biotech programs to develop a diversified funding base that can cushion occasional college or industry budget constraints.
- Advising college programs about industry trends that may require new or modified biotech curricula in order to ensure continuing program relevance to industry needs.
- Aligning with initiatives that have common goals that can leverage funding. Examples include collaboration with the National Center for the Biotechnology Workforce, NBC2, Nano-Link, and Learn to Earn.

Educational and Industry Partnerships

Biotechnology programs at community and technical colleges are intended to prepare students for technical careers in the biotechnology industry. For that reason, when creating educational and industry partnerships, it is critical to address the needs of industry in the educational programs. The 2009 *Replicating Success: Innovative Collaborations Between the Biotechnology Industry and Education* document provides insights into the creation and sustainability of such partnerships.[xxvii] Another example that demonstrates the value of partnerships is described in *Growing Talent: Meeting the Evolving Needs of the Massachusetts Life Sciences Industry.*[xxviii] In addition, biotechnology programs are connecting some their pedagogy with the recommendations of *Vision and Change in Undergraduate Biology Education: A Call to Action.*[xxix]

Need for Career Information and Exploration

The dynamic nature of biotechnology and the rapidly changing advances make it difficult for students to understand the many and varied opportunities that are available in the technical biotechnology workforce. Whereas there is an abundance of information about high-end science careers that require a PhD or MD, there is a lack of information about technical careers in the life sciences other than in health care fields. Bio-Link, in conjunction with other career sites, is working to change that by creating a career site that focuses on the technical careers. The Biotechnology Industry Organization (BIO) has current information available on its website at http://www.bio.org. But more focused career pathway portals help students, teachers, counselors, and parents with the much-needed information about the varied and vast number of biotechnology career possibilities requiring differing levels of preparation. Key to gaining entrance into the biotechnology workforce are soft skills, laboratory and manufacturing competencies, and specific skills. The industry-driven skill-standards projects have informed program development of curricula that provide actual skills development. Links to the skill standards can be found at the Bio-Link website at http://www.bio-link.org. Another excellent resource is *Careers in Biotechnology: A Counselor's Guide to the Best Jobs in the United States, 3rd Edition.*[xxx]

Connections that Encourage Collaboration

One of the most significant career-pathway strategies is to minimize barriers of articulation and collaboration between K–12 institutions, community colleges, and universities themselves as educational institutions, and then to permit partnerships with business and industry, as well as community based organizations and governmental groups. Bio-Link has discovered some working relationships that foster this collaboration. One model is the Bay Area Biotechnology Education Consortium that provides inspiration, professional development, and support for teachers.[xxxi] Oklahoma also has a strong biotech support system for teachers. Other states, such as Georgia, have adopted a biotechnology curriculum that engages students and parents with a renewed interest in science.[xxxii] The Carnegie Institution in Washington, DC, has a long history of reaching out to DC schools with support for biotechnology education. Individuals have developed some outstanding high school materials. Notably, Ellyn Daugherty has developed four years of high school biotechnology curriculum. She has also written a textbook and has conducted teacher professional development workshops around the nation.[xxxiii] Bio-Rad has been a strong supporter of education and has contributed hours of time and made multiple contributions to biotech teacher workshops. George Cachianes, at Abraham Lincoln High School in San Francisco, has written materials. And so it goes with many more examples that could be cited. Every effort needs to be made to make connections with high school programs and community colleges and to promote collaboration.

Some states have adopted the Project Lead the Way (PLTW) Biomedical Sciences (BMS) Program, which includes a sequence of courses aligned with appropriate national learning standards as well as a hands-on, real-world problem-solving approach to learning.[xxxiv] This program offers an appropriate pathway for students exploring careers related to human medicine, but it does not fit well with laboratory and manufacturing biotechnology careers, nor do the courses articulate well with community college biotechnology courses.

Biotechnology Degrees and Certificates

Community colleges offer a variety of degrees and certificates that prepare students for work in biotechnology companies and research institutions. Some of these programs are designed for recent high school graduates; others require a bachelor's degree. Samples of degrees and certificates are listed below and are described in detail at the Bio-Link website.[xxxv]

Examples of Degrees Offered at Select Community and Technical Colleges:
- Biotechnology Associates Degree
- Biomanufacturing Associates Degree
- Bioprocess Laboratory Technology
- Regulatory Affairs Associate Degree
- Agricultural/Biofuels Process Technology

Examples of Certificates Offered at Select Community and Technical Colleges:
- Biotechnician/Biotechnology/Bioscience Certificate
- Applied Biotechnology Certificate
- Bioinformatics Certificate
- Biomanufacturing Certificate
- Medical Devices Certificate
- Quality Control Certificate

- Post-Baccalaureate Certificate/Advanced Technical Certificate
- Post-Baccalaureate Intensive Certificate
- Clinical Research Professional
- Regulatory Affairs Certificate
- Environmental Laboratory Technologist Certificate
- Regulatory Assurance Technologist Certificate
- Genomics Technology Certificate
- Stem Cell Certificate

Typical Program Descriptions
The Biotechnology Laboratory Technician Programs prepare students for work in laboratories involved in research and development, testing, forensics, and quality control. Biomanufacturing Programs prepare students for work in production facilities. Programs provide foundation courses in math and science disciplines including algebra, statistics, chemistry, biology, microbiology, and computer science. Most of the programs incorporate industry internships and career exploration. All the programs are industry driven and include hands-on laboratories that correlate with the skills needed by industry.

Program Outcomes
Students who successfully complete a laboratory technician program will be able to, under the direction of a scientist:
1. Manage laboratory activities including safety, troubleshooting, record keeping, ordering supplies, maintaining equipment, and preparing reports and presentations.
2. Perform technical procedures such as solution and media preparation, sterile and aseptic techniques, DNA extraction, transformation, transfection, electrophoresis, polymerase chain reaction (PCR), enzyme-linked immunosorbent assay (ELISA), Western blots, maintenance of cell lines, protein isolation and purification using chromatographic techniques, and cell labeling and counting.
3. Conduct experiments following standard operating and safety protocols.
4. Analyze data using computer technology including maintaining databases, preparing spreadsheets, conducting statistical

analysis, utilizing bioinformatics and presenting graphical displays.

Students who have training in biomanufacturing will be able to work in production facilities and clean rooms that use standard operating procedures (SOPs).

Examples of Biotechnician Positions

The titles of the biotechnicians vary. Some examples include laboratory assistant, lab specialist, research assistant, and manufacturing technician. Employers include university and commercial biotechnology research and production labs, pharmaceutical labs, manufacturing facilities, forensic labs, fisheries, and natural resource management organizations. Entry-level salaries vary across the country, and benefits are generally attractive and include incentives for additional education.

Concluding Remarks

The biotechnology and related life-science industries are strong. The need for skilled technicians is evident, and community colleges across the nation are responding with programs based on industry input. The challenges of developmental math, articulation between educational institutions, and availability of appropriate career information leap to the forefront of discussions about biotechnician education. But the bottom line seems to be the need to engage graduates who are now working in the industry. They are the ones who have benefited from competency-based skills education and have experienced entry into the industry and promotions within companies. They provide the strongest and most compelling voices that can attest to the fact that there are opportunities for the "neglected majority" and the need for middle-skill jobs. In fact, one might question the term "middle skill" since, once hired into companies, many graduates become master technicians and managers. Our next step in fixing a broken educational system is to engage the expertise and insight of the products (now employees) of the biotech programs at community and technical colleges.

[i] BayBio, *Translating science into better health: Impact 2008* (San Francisco, CA: BayBio, 2008).

[ii] BayBio, *The Future is Here: A Challenge for California's Leadership* (San Francisco, CA: BayBio, 2010).

[iii] Battelle, *Technology, Talent and Capital: State Bioscience Initiatives,* 2008.

[iv] Burrill & Company, *Life Sciences: A 20/20 vision to 2020,* 22nd Annual Report on the Industry (San Francisco, CA: Burrill & Company, 2008).

[v] Battelle, *Battelle Annual Report,* 2006.

[vi] Gus Koehler and Victoria Koehler-Jones, *California's Biotechnology Workforce Training Needs for the 21st Century* (Sacramento, CA: Applied Biological Technologies Initiative Economic and Workforce Development Program, California Community Colleges, 2006).

[vii] U.S. Bureau of Labor Statistics, National Employment Data, available at http://www.dol.gov (accessed 2011).

[viii] North Carolina Biotechnology Center, "Growing Biotechnology Statewide," http://www.ncbiotech.org (accessed April 20, 2010).

[ix] Elaine Johnson, "Crossroads to the Future," *Biolink Connection* 10, no. 2 (April 2009): 2.

[x] Kristin Hersbell Charles and Elaine A. Johnson, *National Trends in Biotechnology Technician Education, Bio-Link Survey Analysis, Brief Update 2008* (Bio-Link, 2008).

[xi] Information found through http://www.ncbiotech.org (accessed 2010).

[xii] Information about Shoreline's biotechnology program can be found at http://new.shoreline.edu/biotechnology/default.aspx.

[xiii] BayBio, *The Future is Here.*

[xiv] American Association of Community Colleges, "Fast Facts," http://www.aacc.nche.edu/AboutCC/Pages/fastfacts.aspx (accessed April 20, 2012).

[xv] Elaine A. Johnson, "Bio-Link: Educating the Biotechnology Workforce Using Resources of Community and Technical Colleges," *Biochemistry and Molecular Biology Education* 31, no. 5 (2003):348–351.

[xvi] Julian L. Alssid, Melissa Goldberg and John Schneider, *The Case for Community Colleges: Aligning Higher Education and Workforce Needs in Massachusetts* (The Boston Foundation, 2011).

[xvii] Madeline Patton, *Educating Biotechnicians for Future Industry Needs* (Washington, DC: Community College Press, 2008)

[xviii] Beata D. Kochut and Jeffrey M. Humphreys, *Shaping Infinity: The Georgia Life Sciences Industry Analysis 2011* (Selig Center for Economic Growth, 2011).

[xix] Battelle, *Taking the Pulse of Bioscience Education in America: A State-by-State Analysis* (A Report Prepared by Battelle in Cooperation with Biotechnology Industry Organization and the Biotechnology Institute, 2009).

[xx] Federal Register, Office of Science and Technology Policy, October 11, 2011.

[xxi] *Georgia Bio E-Chronicle*, "HS Biotech Course Boosts Student, Parent Interest in Science," July, 2010, http://www.georgiabioed.org (accessed April 20, 2012).

[xxii] Katy Korsmeyer, *Bay Area Biotechnology Education Consortium: A Model Organization for Inspiring and Supporting High School Biotechnology Teachers* (Bio-Link, 2010).

[xxiii] National Career Pathways Network and the Institute for a Competitive Workforce, *Thriving in Challenging Times: Connecting Education to Economic development through Career Pathways* (2009) http://www.ncpn.info/thriving-in-challenging-times.php (accessed April 20, 2012).

[xxiv] Information found through http://www.ncbiotech.org (accessed 2010)

[xxv] Lisa A. Seidman, *Basic Laboratory Calculations for Biotechnology* (Benjamin Cummings, 2007); Lisa A. Seidman and Cynthia J. Moore, *Basic Laboratory Methods for Biotechnology, 2nd edition* (Benjamin Cummings, 2008); Lisa A. Seidman, Mary Ellen Kraus, Diana Brandner, and Jeanette Mowery, *Laboratory Manual for Biotechnology and Laboratory Science: The Basics* (Benjamin Cummings, 2010).

[xxvi] John Carrese, *Equipment Depot Resource Guide* (Bio-Link, 2011).

[xxvii] Lori Lindberg, *Replicating Success: Innovative Collaborations Between the Biotechnology Industry and Education* (Bio-Link, 2009).

[xxviii] UMass Donahue Institute, Massachusetts Life Sciences Center, and Massachusetts Biotechnology Council, *Growing Talent: Meeting the Evolving Needs of the Massachusetts Life Sciences Industry*, 2008.

[xxix] *Vision and Change in Undergraduate Biology Education: A Call to Action,* A Summary of Recommendations made at a National Conference Organized by the American Association for the Advancement of Science with Support from the National Science Foundation, 2009.

[xxx] Gina Frierman-Hunt and Julie Solberg, *Careers in Biotechnology: A Counselor's Guide to the Best Jobs in the United States, 3rd ed.* (California Applied Biotechnology Centers and Hubs of the California Community Colleges Economic Workforce Development Program and Bio-Link, 2008).

[xxxi] Korsmeyer, *Bay Area Biotechnology Education Consortium.*

[xxxii] *Georgia Bio E-Chronicle*, "HS Biotech Course Boosts."

[xxxiii] Information is available at http://www.bioteched.com.

[xxxiv] See http://www.pltw.org.

[xxxv] http://www.bio-link.org.

10 Chapter

Secondary/Postsecondary Programs of Study for Nanotechnology Technicians

Deb Newberry
Nano-Link: NSF/ATE Center for Nanotechnology Education
Dakota County Technical College, MN

Introduction

Nanotechnology has meant many things to many people over the last two or three decades. It has meant a solution to global problems, disastrous environmental situations, the new dot-com economic driver, and a myriad of amazing career opportunities. People are working to provide nanotechnology-based solutions for global problems, and many are working to ensure that nanoscience disasters do not occur. The last decade has most likely proven that nanotechnology is an arena for patient capital rather than for get-rich-quick dreams. What nanotechnology has proven to be is an area with diverse career opportunities and options for those just entering the workforce or for those who have been employed for many years. For example, paint companies are using nanoparticles to increase the ease of application and wearability of paints. Golf clubs and tennis rackets are using nanofibers to increase flexibility and strength while making the sport equipment lighter in weight. And nanoscale particles are used in pregnancy and drug-detection kits as well as cosmetics.

Nanotechnology is proving to be a technology that has or will have a significant impact on every industrial segment—from communication to construction, from medical devices to automobile manufacturing, and from disease diagnosis to water purification. This chapter will discuss what nanotechnology is, how it is being applied to various industries, and related educational requirements and options, job opportunities, and career paths.

What is nanotechnology?

Many are familiar with the prefixes "mega," meaning a million, and "giga," meaning a billion, because of the ever increasing density of memory devices. Similar Greek prefixes represent smaller amounts: for example, "centi-" means one-hundreth and "micro" means one-millionth. We are familiar with those prefixes from terms such as centimeter, microbe, and microelectronics. Similarly, nanoscience or nanotechnology is, in the purest sense, the application of a prefix—*nano*—to the nouns *science* and *technology*. *Nano* is the Greek word for dwarf and is the prefix used to designate one-billionth. So a nanosecond is one-billionth of a second, a nanogallon is one-billionth of a gallon, and a nanometer is one-billionth of a meter. In essence, it is a very small amount of anything: time, volume, or distance, for example.

In common use today, nanotechnology most often uses the prefix nano as a modifier that refers to length. The common unit is the nanometer and is abbreviated "nm." A human hair is 100,000 nm in diameter, a red blood cell is 2000 nm by 7000 nm, and ten hydrogen atoms lined up in a row would be one nanometer long. When we are working at the nanoscale, we are working at the molecular and atomic level. Figure 10.1 from the U.S. Department of Energy shows a length scale in the center with natural objects along the left-hand side and manmade objects of the same size scale on the right-hand side. As the figure shows, various researchers have been able to replicate the nanoscale of nature in the laboratory.

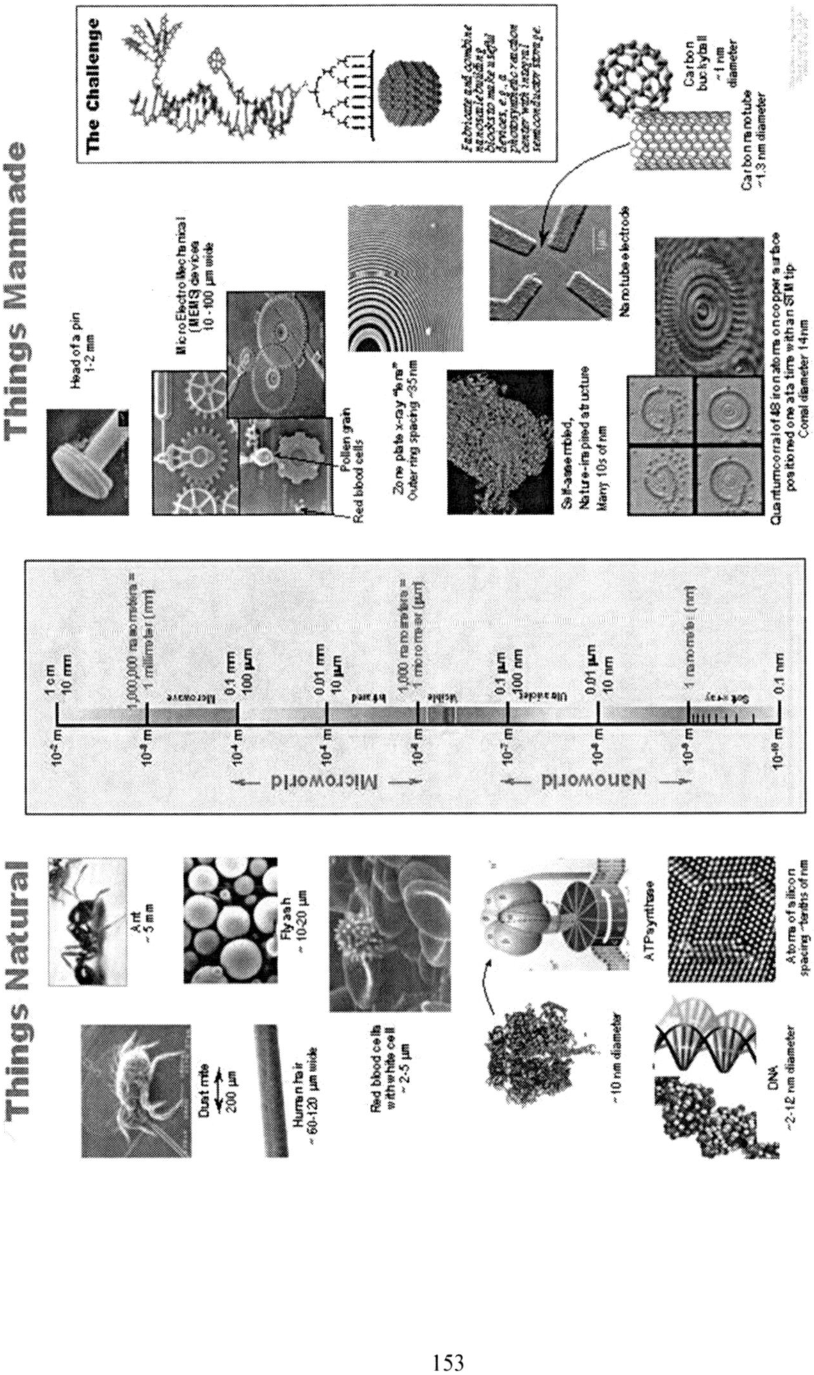

Figure 10.1. The Scale of Things – Nano meters and More Courtesy of the Office of Basic Energy Sciences, Office of Science, U.S. Department of Energy

A good definition for nanotechnology is: *Nanotechnology is the ability to observe, image, study, measure, and manipulate at the molecular and atomic scale.*

Even though molecules can be thousands of nanometers in certain dimensions, the interactions that govern their properties often occur over distances of a few tens of nanometers, and small groups of atoms certainly fall in the "less that 100 nm" range that is often used to define nanotechnology.

Nanotechnology is important because the atomic and molecular structure of every material defines the physical, electrical, and biological properties of that material. Figure 10.2 shows this relationship. A classic example starts with the element carbon. Taking carbon atoms and exposing them to certain temperatures and pressures results in charcoal. Exposing those very same carbon atoms to different temperatures and pressures can cause a different arrangement of the atoms and result in diamond. Both of these materials are made from the same atom, carbon, but because the atoms are arranged differently, the two resulting materials certainly have different properties (and value).

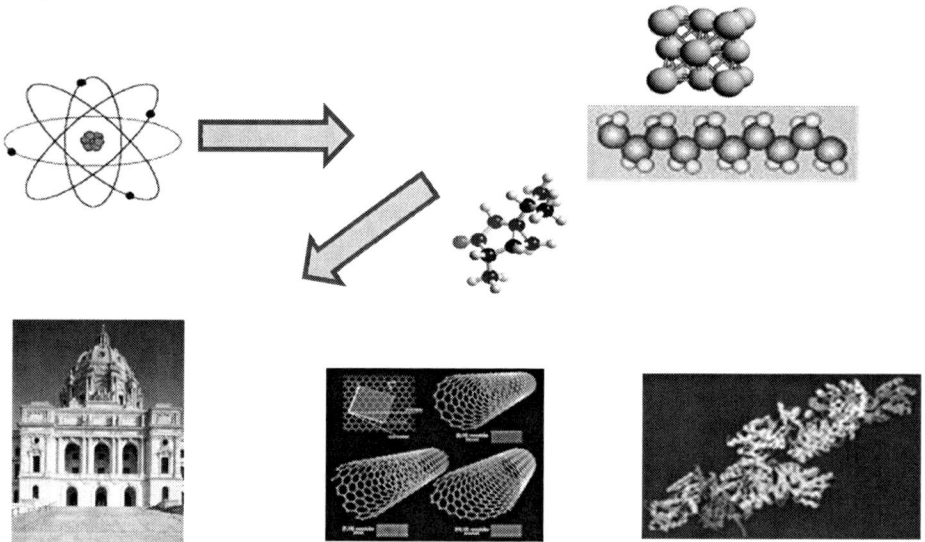

Figure 10.2. Atoms form molecules, and the arrangement of those atoms and molecules defines the physical, electrical, and biological properties of materials.

154

Because atoms and molecules are fundamental to all materials and interactions, people who work in any of the traditional sciences are required to study and understand phenomena at the nanoscale (see figure 10.3).

Aspects of physics, chemistry, biology and material science all require an understanding of nanoscience. For example, scientists in these fields might need to understand the complex interactions of proteins, composite materials, the immune system, electrical circuits and structures such as carbon nanotubes, or the interaction between the foot of a gecko and glass. And when scientists in these fields study phenomena at the nanoscale, knowledge or deeper understanding is added to the traditional disciplines; the influence goes in both directions.

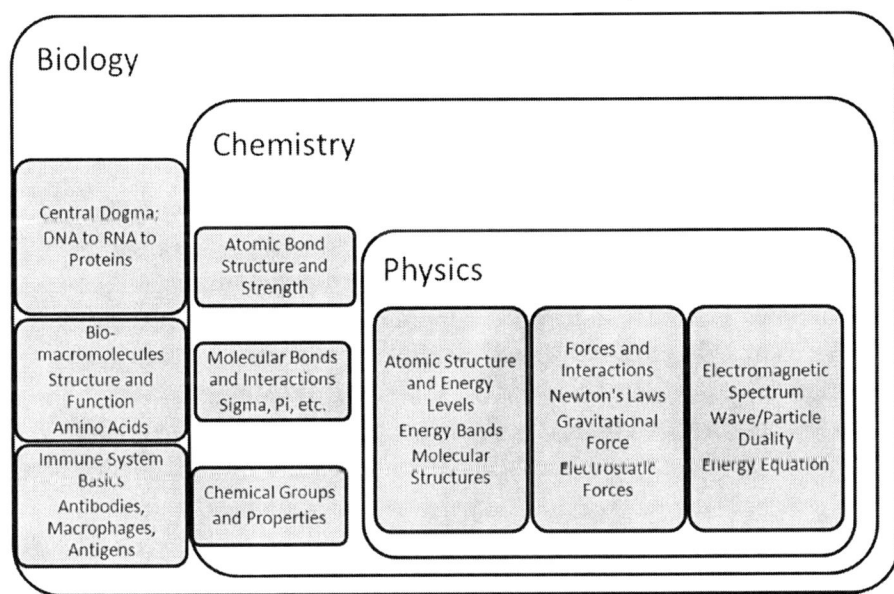

Figure 10.3. Certain aspects of traditional science require an understanding of the nanoscale.

Each of the traditional disciplines builds upon and is integrated with the others at the nanoscale. For example, to understand nanoscale interactions of biological systems, it's necessary to understand some concepts from physics and some from chemistry, such as atomic and molecular bonds. Pieces of each of the traditional sciences need to be understood and integrated to describe the world at the nanoscale.

Nanotechnology has a significant, multidisciplinary, and bidirectional relationship with the other sciences. It both requires concepts from all the traditional disciplines and provides information that contributes back to those disciplines. Because of this interrelationship with the other sciences, nanotechnology is applicable to a multitude of industries. Our understanding of nanotechnology is helping us understand friction, how blood moves through capillaries, and how drugs interact with proteins. As a result of our new understanding of the nanoscale, better and longer lasting lubricants are being developed, artificial tissues and organs may be possible, and treatment drugs may result in fewer side effects. Nanotechnology is influencing paints, lubricants, and material coatings; it is impacting electrical devices and computer design. Nanotechnology is changing medical diagnostics by making it possible to use smaller sample sizes and to test for multiple diseases at once. In battery technology and solar panels, nanotechnology is improving efficiency while making it possible to use less-expensive materials. And nanotechnology is providing sensor systems with greater specificity and sensitivity.

In addition, because of nanotechnology and our developing understanding of nanoscale phenomena, the lines of definition between the sciences and various market segments are becoming blurred. In many cases, companies have interdisciplinary teams to create new products and technologies as well as improve existing products. Educational institutions and students are realizing that a "stovepipe approach" no longer works. Figure 10.4 shows these three phenomena:

1. Nanotechnology is influencing and is influenced by the traditional sciences,
2. Nanotechnology can affect multiple markets, and
3. Nanotechnology is blurring the lines of distinction between sciences and various industries.

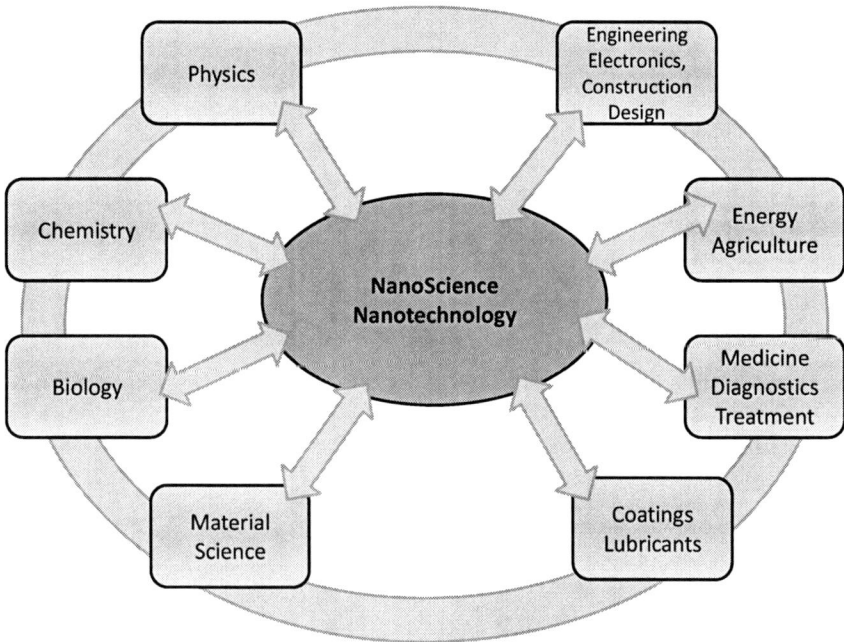

Figure 10.4. The inter- and multidisciplinary aspects of nanotechnology.

A note about nanoscience and nanotechnology: There is indeed a distinction between science and technology. Science can be viewed as the development of understanding or the discovery of the fundamental interactions of the world around us. Science often is synonymous with research—both focus on discovery in the absence of an intended outcome or application. Technology can be associated with engineering—the application of scientific discoveries or the products developed from scientific discoveries. Based on this understanding of science and technology, *nanoscience* can be thought of as the observation and study of nanoscale objects and phenomena, and *nanotechnology* as the application of that discovered understanding to materials, product development, or processes.

Nanotechnology Technician Supply and Demand

Because nano-savvy employees are and will continue to be needed in a multiplicity of industries, the number of careers and job opportunities in this field is great. Mihail Roco, Senior Advisor for Nanotechnology at the National Science Foundation, states that the United States will need over eight hundred thousand nanoscience technicians in the next decade.[i] The number of education programs that focus on or include nanoscience continues to increase, and the need for skilled employees is significant. In Minnesota, nanotechnology program graduates are getting jobs, and industries are requesting more graduates. In companies, PhD employees are doing nanotechnology technician work because they alone have the expertise to prepare samples, operate the equipment, and analyze results.

As the study and application of nanotechnology has matured over the last several decades (figure 10.5), the requirements for workers in the area have changed.

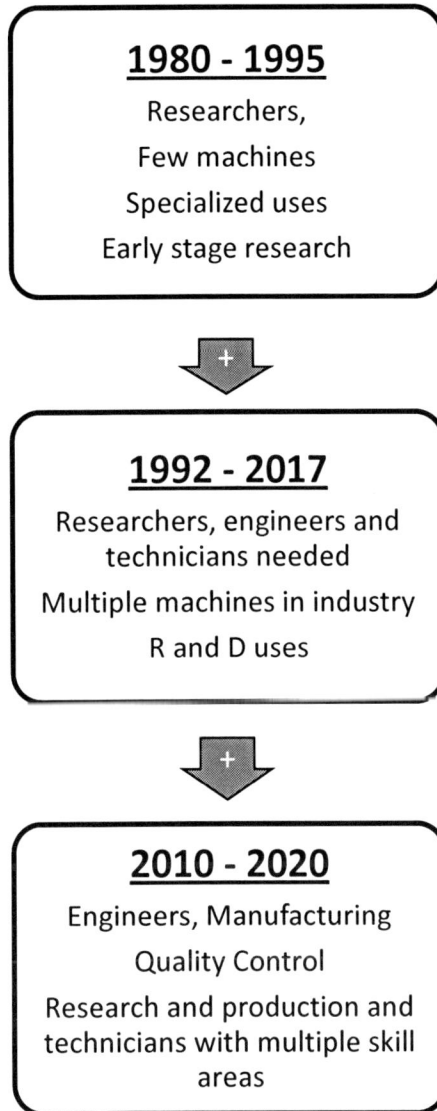

Figure 10.5. The application of nanoscale phenomena has changed over time and will continue to change, driving a need for engineers, scientists, and technicians with multiple skills, knowledge, and abilities in nanoscience and nanotechnology.

In the 1980s, only a few researchers at universities or companies had access to equipment that could image or manipulate individual atoms, and only a few people knew how to operate or use this equipment. The scanning tunneling microscope (STM) was invented in 1981, and the atomic force microscope—now the workhorse of nanotechnology—was invented in 1986. At that point in time, very few people were involved in work at the nanoscale, and those involved were experienced, creative PhDs. The developers of the STM, Gerd Binnig and Heinrich Rohrer, were awarded the Nobel Prize in Physics in 1986.

During the early 1990s, the electronic-device industry began to edge closer to creating transistors that were entering the "hundreds of nanometers" range—technology was getting smaller and smaller to provide more computing power in each square centimeter. Because of this desired reduction in size, the electronics industry began to look for employees that understood the nanoscale and apply this understanding to design and manufacturing. Initially, employees working at the nanoscale were engineers and scientists. As the manufacturing process was refined, the industry increasingly needed operation technicians who had training in the new equipment used to produce nanoscale transistors. In the late 1990s, Pennsylvania State University created the first two-year nanoscience technician program to meet the needs of the electronics industry.

In the early 2000s, other industries (see figure 10.6) were becoming aware of the various properties of nanoscale materials and coatings, as well as the ever-increasing cadre of tools that could be used to create and characterize materials at the nanoscale. In early 2003, Dakota County Technical College (DCTC) held an industry symposium to evaluate the need for nanoscience technicians. Representatives attended from a significant number of area companies in the material, electronics, and biotechnology industries. At that point, DCTC decided to create a multidisciplinary nanoscience technician program to meet the needs of area industries.

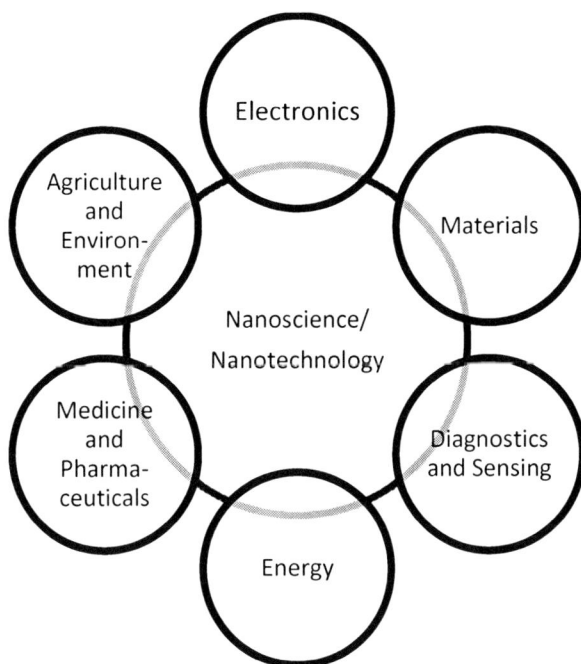

Figure 10.6. Today, nanotechnology is affecting virtually every industry in existence. As nanoscale discoveries unfold and are applied, new companies and industry segments will be created to develop electrically conductive composite materials, to create energy using vibrational motion, and more.

Future trends for nanotechnology careers will expand beyond research and product development into manufacturing, production, quality assurance, and testing. In addition, new discoveries, equipment modifications, and greater understanding of material and interactions at the nanoscale will all continue to fuel development at an increasing rate. This will result in more opportunities for improved and new products, as shown in figure 10.7.

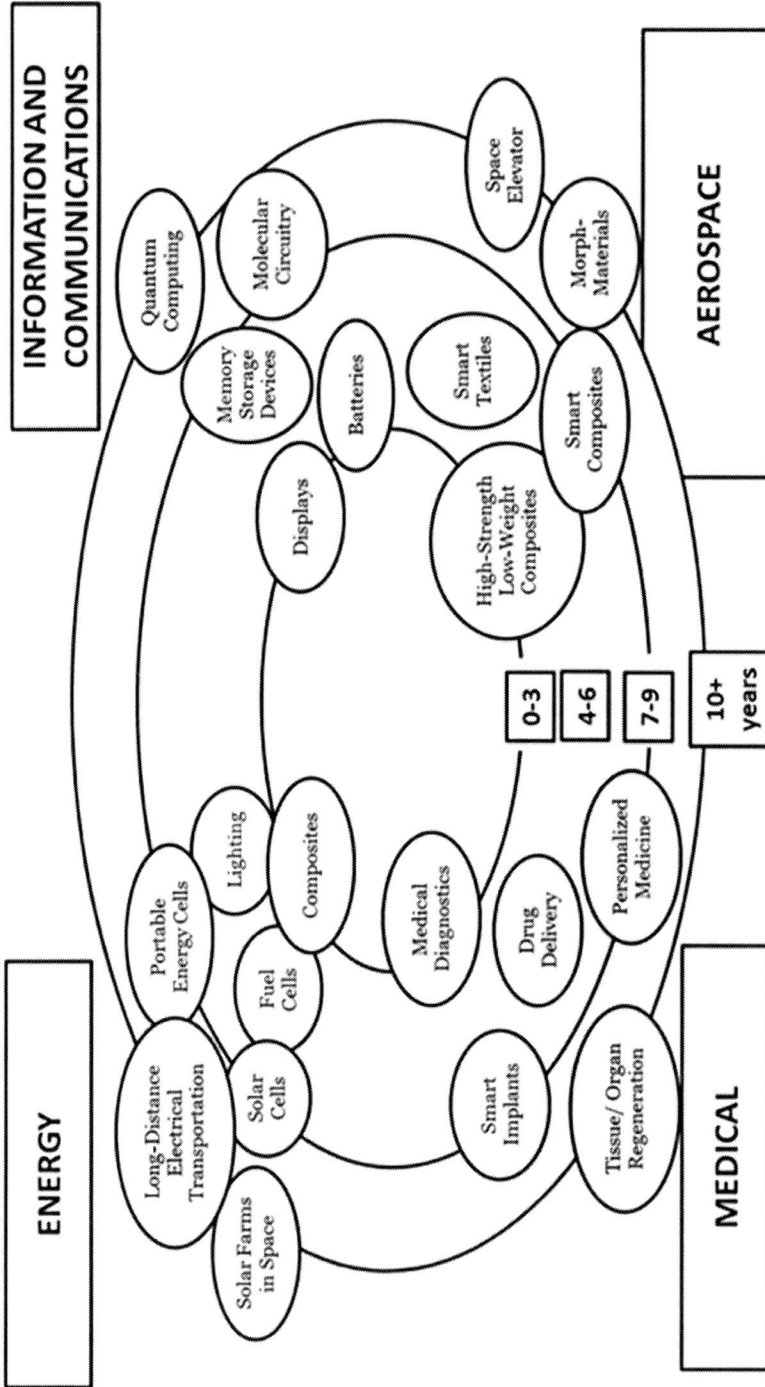

Figure 10.7. New discoveries and applications are anticipated for nanotechnology (source: Rice University).

A Nanotechnology Program

As mentioned above, in 2003, Dakota County Technical College was evaluating the possibility of creating a two-year nanoscience technologist program. The college's questions to industry were: "If nano-savvy technicians were graduated, would you hire them?" and "What are the specific skills, knowledge, and abilities that you would like these employees to have?" The answer to the first question was a resounding "yes." The answer to the second question took longer to develop.

DCTC is located in the southern portion of the Minneapolis/St. Paul area. The Twin Cities are home to companies such as 3M, HB Fuller, Valspar Paints, Medtronic, Boston Scientific, R and D Systems, Minnesota Wire and Cable, Seagate, Honeywell, and Cypress Semiconductors. This wealth of industries also covers a diverse set of applications. Based on conversations with various industry groups and individuals, DCTC decided to create a multidisciplinary program that would include education in nanomaterials, nanobiotechnology, and nanoelectronics to meet the needs of area industries. The resulting program is shown in figure 10.8. The two-year AAS degree program consists of seventy-two credits, of which forty-three are nanotechnology-specific courses. This program is based on a strong partnership with the University of Minnesota. In the fourth semester, the DCTC students attend the university for lectures and lab classes in the three focus areas—materials characterizations, electronics, and biotechnology.

Semester 1 at DCTC			Semester 2 at DCTC			Semester 3 at DCTC			Semester 4 At Univ. of MN		
Course	Name	Credits	Course	Name	Credits	Course	Name	Credits	Course	Name	Credits
BIOL 1500	General Biology	4	CHEM 1500	Introduction to Chemistry	4	NANO 2101	Nano Electronics	3	MT 3111	Elem. of Micro Manufacturing	3
PHYS 1100	College Physics I	4	PHYS 1200	College Physics II	4	NANO 2111	Nanobiotechnology/ Agriculture	3	MT 3112	Elem. of Micro Mfg Lab	1
			NANO 1222	Student Lab Experience and Research	3	NANO 2121	Nanomaterials	3	MT 3121	Thin Films Deposition	3
ENGL 1100	Writing & Research Skills	3	SPEE 1020	Interpersonal Communication	3	NANO 2131	Manufacturing, Quality Assurance	2	MT 3131	Intro to Materials Characterization	3
MATS 1300	College Algebra	4	MATS 1250	Principles of Statistical Analysis	4	NANO 2140	Interdisciplinary Lab	3	MT 3132	Materials Characterization Lab	3
NANO 1100	Fund. of Nano I	3	NANO 1200	Fund of Nano II	3	NANO 2151	Career Planning and Industry	1	MT 3141	Principles and Applications of Bionanotechnology	3
					1				MT 3142	Nanoparticles & Biotechnology Lab	1

DAKOTA COUNTY
TECHNICAL COLLEGE

Figure 10.8. The AAS-Degree Nanoscience Technologist program at Dakota County Technical College. At least eight U.S. colleges have adopted courses and larger segments of this curriculum.

164

The creation of the program, courses, and specific course content was and continues to be an iterative process between educators from DCTC and the University of Minnesota and area industry representatives.

At the onset of curriculum development, industry representatives stated that they wanted employees who: 1) understood how to use the tools of nanoscience, 2) understood the concepts involved in working with nanoscale objects and phenomena, 3) were skilled in critical thinking and analytical processes, and 4) could work well in teams. Figure 10.9 shows this set of requirements.

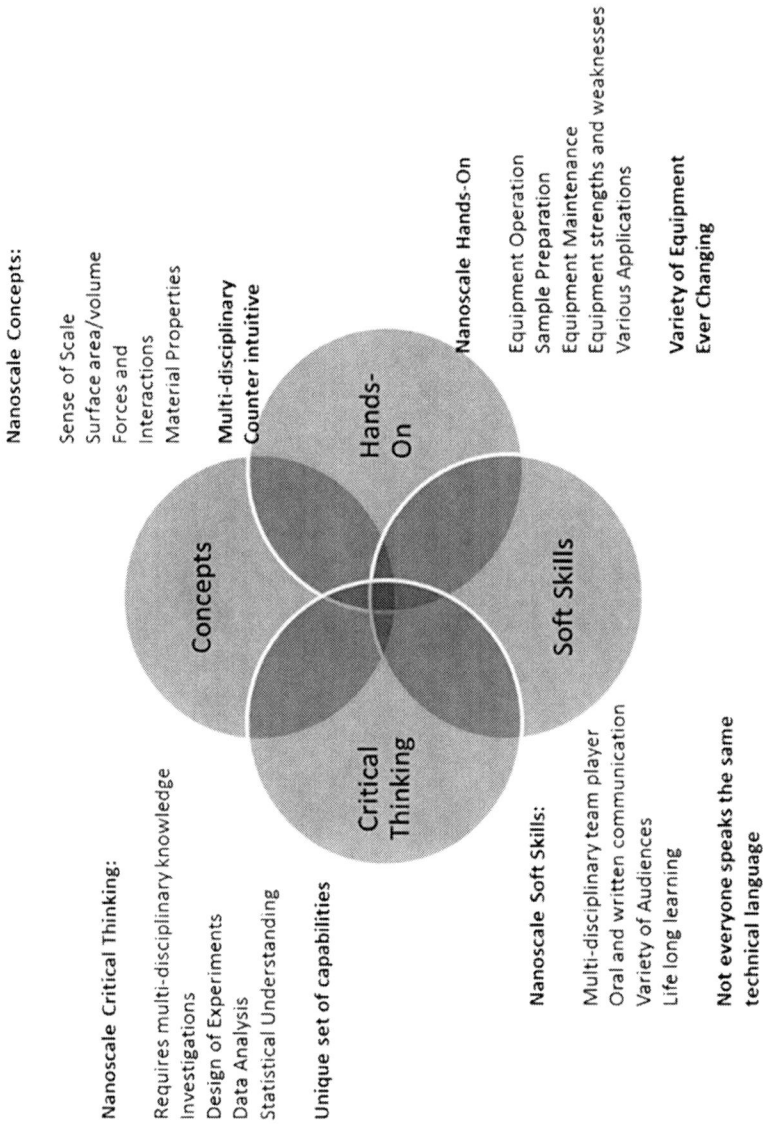

Nanoscale Concepts:

Sense of Scale
Surface area/volume
Forces and
Interactions
Material Properties

Multi-disciplinary
Counter intuitive

Nanoscale Hands-On

Equipment Operation
Sample Preparation
Equipment Maintenance
Equipment strengths and weaknesses
Various Applications

Variety of Equipment
Ever Changing

Nanoscale Critical Thinking:

Requires multi-disciplinary knowledge
Investigations
Design of Experiments
Data Analysis
Statistical Understanding

Unique set of capabilities

Nanoscale Soft Skills:

Multi-disciplinary team player
Oral and written communication
Variety of Audiences
Life long learning

Not everyone speaks the same
technical language

Figure 10.9. Industry representatives provided a list of desired nano-savvy employee skills, knowledge, and abilities.

Based on industry input, DCTC developed a program that encompassed all the desired outcomes. The first year concentrates on nanoscale concepts, independent of specific industry disciplines, similar to the content shown in figure 10.3. Students are also introduced to the tools of nanoscience: atomic force microscopes, scanning electron microscopes, Raman spectroscopy, scanning tunneling microscopes, etc. This introduction includes basic principles of operation as well as sample preparation and procedures. The second year of the program delves deeper into concepts and equipment, and emphasizes critical thinking and analysis. Figure 10.10 shows the basic construction of program content.

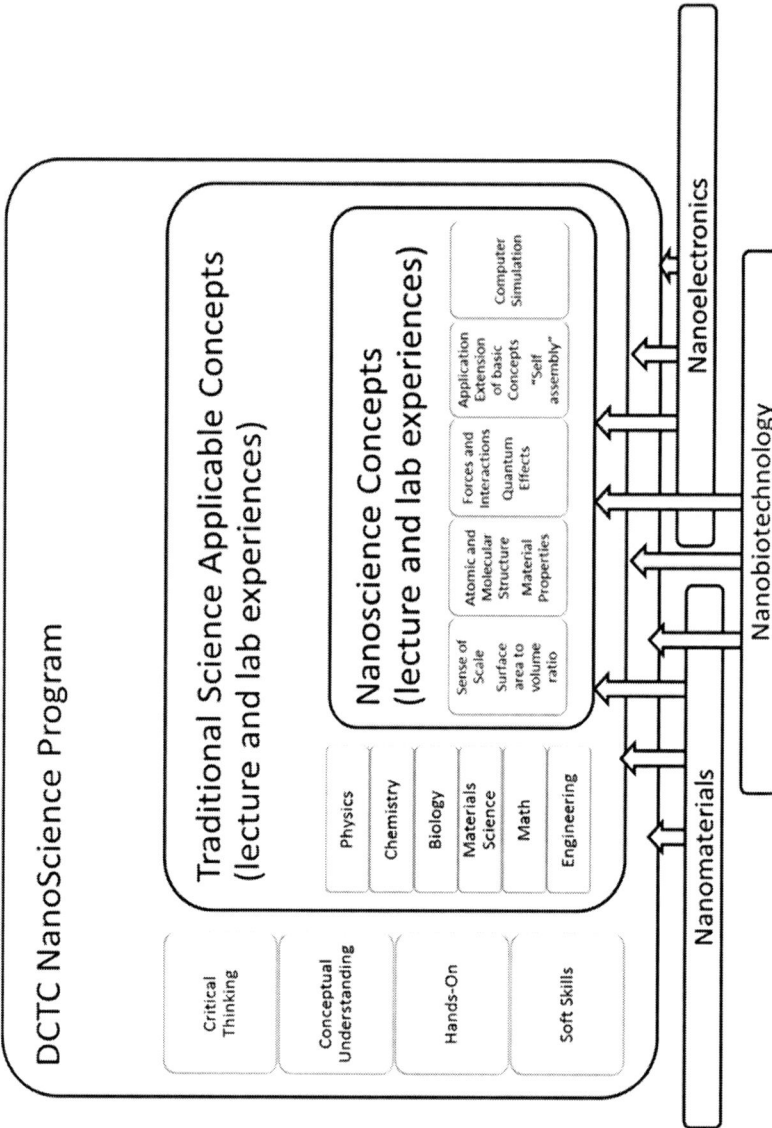

Figure 10.10. Integration of traditional, nanoscience, and industry concepts and requirements into the two-year nanoscience technologist program.

Jobs and Career Paths

Finding a job in nanotechnology can be both very difficult and very easy. First, because of the newness of nanotechnology, many companies do not yet have a job category or title that includes the word "nanotechnologist" or "nanoscientist." Because of this, a search of the Internet or the newspaper want ads may come up with few opportunities for a job with "nano" in the title.

On the other hand, because nanoscience can be applied to so many different disciplines, jobs that require knowledge and skills at the nanoscale are becoming increasingly easier to find.

Figure 10.11 shows how to identify some of the titles for jobs that include nanotechnology work. For example, if you select one word from the "Category Description" column, one word from the "Task Arena" column, and one word from the "Human Resource Descriptor" column, you will have created one of the myriad of job titles being used for the graduates of nanoscience technician programs. Titles that result include Electronic Design Technician, Biotech Laboratory Aide, and Medical Device Test Technician.

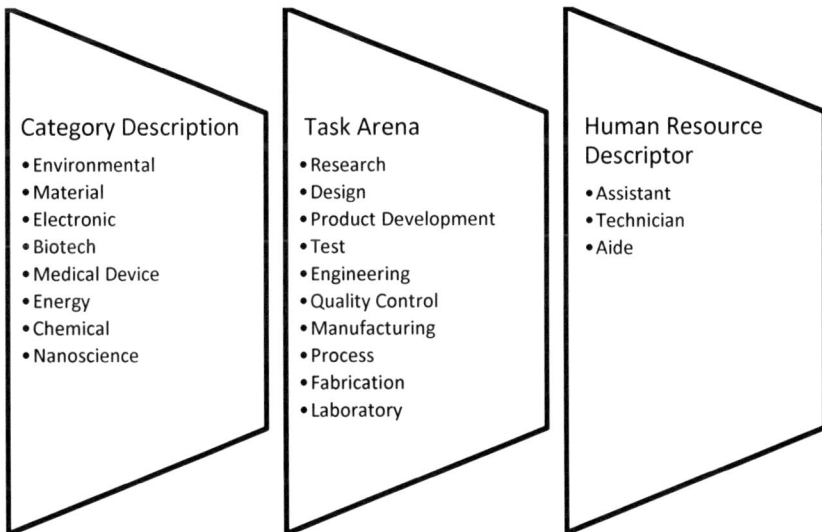

Category Description	Task Arena	Human Resource Descriptor
• Environmental	• Research	• Assistant
• Material	• Design	• Technician
• Electronic	• Product Development	• Aide
• Biotech	• Test	
• Medical Device	• Engineering	
• Energy	• Quality Control	
• Chemical	• Manufacturing	
• Nanoscience	• Process	
	• Fabrication	
	• Laboratory	

Figure 10.11. Multiple job titles are possible in nanoscience and nanotechnology careers.

Currently, many nanotechnology jobs support basic research or product development. Technicians usually work as part of a team composed of scientists, engineers, and other technicians. Their jobs may involve creation of nanoparticles or different coatings involving nanomaterials. Often the job will also involve the testing, analysis, and documentation of test results. Technicians in this type of research-assistant job often comment that the job never stays the same. They are not doing the same thing they were doing six months ago and don't expect to be doing their current task six months in the future. It is a job that requires being on your toes and willing to learn (and think) quickly.

Another career path that is becoming more prevalent for nanoscience technicians is testing, product control, or quality control. As equipment is developed to test products, coatings, or any material at the nanoscale, the requirements for testing have become more stringent. Nanoscience technicians are finding significant opportunities in the medical-device industry testing implantable devices and the coatings that cover those devices. This type of job can be intense when there is a problem that needs to be solved quickly, but these jobs also provide significant opportunities for creativity, the use of investigative skills, and critical thinking. When a problem is solved, technicians may be part of a team of "company heroes." Technicians that work in this area also are anticipatory and work to determine how best to test a product or maintain critical operational parameters. In this way, they serve as proactive participants in the product line.

Another strong job segment for nanoscience technicians is working in start-up or small companies. In these types of work environments, one employee is often expected to do multiple jobs, and a hired technician is no exception. Our increasing understanding of the nanoscale has fueled a significant growth in small start-up companies. These companies may be in the area of nanoparticle research and fabrication, hydrophobic surface coatings, flexible electronics, or superstrong structural materials. In all of these cases, technicians may find themselves using an atomic-force microscope one day, growing and analyzing chemical samples the next, and measuring wear resistance with a nanomechanical measuring system the next. This is clearly a job that takes advantage of the broad reach of nanoscience into traditional sciences and requires a plurality of equipment skills.

Finally, nanotechnology is a STEM career path that can keep expanding. In the fall of 2011, Dakota County Technical College, in partnership with Independent School District 917 in Minnesota, offered the first yearlong, transferrable set of courses in nanotechnology to high school juniors and seniors. Eighteen students are currently in the program, and many of them are planning to include nanotechnology in their future educational plans.

Nanotechnology concepts can be directly integrated with four-year degrees in many areas. (Note that chemists, biologists, and physicists have been doing nanoscience for centuries—it just never had that name.) Various degree programs, especially in science and engineering fields, have begun to integrate nanoscience into courses. Figure 10.12 shows the relationship between courses taught in the DCTC two-year program and nanoscale concepts and content areas in traditional electrical engineering courses.

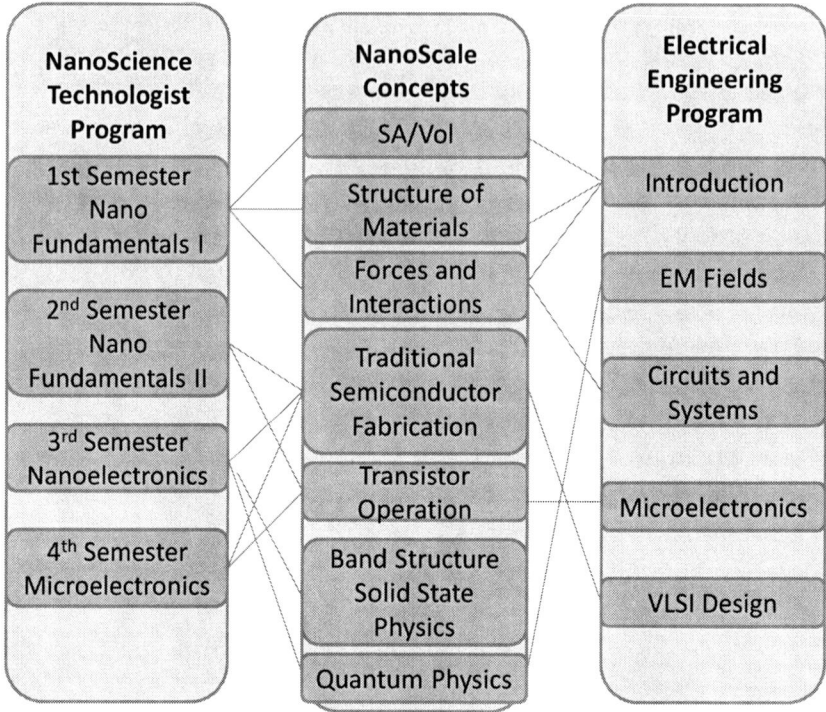

Figure 10.12 Correlation between nanoscience courses, concepts and electrical engineering courses.

4+2 Curriculum Pathway

Students can begin to prepare for a career in nanotechnology in early high school by taking a basic math and science course program. There are a small number of high schools that may include nanotechnology as a segment of a career-planning or pre-engineering course or as a quarter-long stand-alone course, but the majority of high school students will find nanotechnology concepts among those taught in traditional classes.

Figure 10.3 shows appropriate points where nanoscale concepts might be inserted into coursework in traditional disciplines. Nano-Link, an NSF funded center for nanotechnology education (www.nano-link.org) has a good set of modularized hands-on nanotechnology activities designed for incorporation into chemistry, physics, and biology classes. Nano-Link is also one of several centers that offer professional development workshops for high school educators who want to learn about nanotechnology and integrate it into their curriculum. There are numerous colleges and universities that offer summer nanotechnology experiences for high school students, as well.

A nanotechnology survey course that covers the ancient and recent history of nanotechnology, as well as applications, is also an excellent option for high school students. Many programs at postsecondary institutions offer a nanotechnology survey course for engineering students.

Postsecondary nanotechnology programs are as varied as nanotechnology itself. Some programs focus on one aspect of nanotechnology, such as nanoelectronics. Others, such as the program at DCTC, are more multidisciplinary in nature. Figure 10.13 provides a representative set of courses that could support the nanotechnology career path.

Soph 2			Nano photonics	Biofuels	Energy	Nano biotech	Micro electronics Operation and Fabrication
Soph 1	Precal		Disciplinary Lab	Nano materials	Nano electronics		Manufacturing and Quality Assurance
Fresh 2	Statistics		College Chemistry	College Physics II	Student Lab I	Biotech Ag	
Fresh 1	College Algebra	Writing and Research Skills	College Biology	College Physics I		Materials Characterization	
12th Grade	Algebra 2w/Trig	English 12	Gov't	Physics	Health	Intro to Nano biotech	Fundamentals of Nano II
11th Grade	Math App	English 11	American History	Chemistry	Physical Education	Principles of Nanotech	Fundamentals of Nano I
10th Grade	Geometry	English 10	World History	Biology	Physical Education	Foreign Language	Principles of Engineering
9th Grade	Algebra 1	English 9	Geography	General Science	Physical Education	Foreign Language	Intro to Engr Design

Figure 10.13. Model curriculum for a nanotechnology career pathway. Nanoscale concepts can be included in many of the traditional science and math courses. There are multiple paths and disciplines for postsecondary students.

Conclusion

The ability to observe and understand the world at the nanoscale offers unimaginable opportunity. It gives us an opportunity for enhanced richness of understanding of the nonbiological and biological worlds, and an opportunity to replicate these worlds. It also creates opportunities for students interested in STEM subjects and careers. Skill in observing and measuring materials at the molecular and atomic level can be applied to a myriad of disciplines and career paths. Nanoscience is an arena that will only expand over the coming decades.

---

[i] M. C. Roco, "Converging Science and Technology at the Nanoscale: Opportunities for Education and Training," *Nature Biotechnology* 21, no. 10 (2003): 1247–49.

Secondary/Postsecondary Programs of Study for Microsystems Technicians

Matthias W. Pleil, Ph.D.
Southwest Center for Microsystems Education
University of New Mexico

What are MEMS and Microsystems?

Microelectromechanical systems (MEMS) are very small devices or groups of devices that integrate both mechanical and electrical components. MEMS are typically constructed on one chip that contains one or more micro-components and the electrical circuitry for inputs and outputs. The components include different types of transducers (sensors and actuators), electronics, and structures (e.g. gears, tilt mirrors, diaphragms). Each type of component is designed to interface with an input such as motion, light, fluids, gases, electromagnetic radiation, pressure, temperature, or biomolecules.

MEMS are found in a wide range of biotechnology, transportation, homeland security, and consumer product applications. Common examples include crash-bag sensor systems; ink-jet print heads; Digital Light Processing (DLP) television and projection systems; microphones; and motion sensors found in smart phones, remote controls, biometric, and game controllers. The terms "MEMS" and "Microsystems" are used interchangeably; in Europe, it is most common to use the term Microsystems Technology (MST), while in North America, "MEMS" is more commonly used.

The Southwest Center for Microsystems Education (SCME) web site (www.scme-nm.org) has a series of educational materials including MEMS Applications, MEMS History, and BioMEMS learning modules that have summaries of the more prevalent MEMS devices found in use today.

A MEMS Industry Perspective: Motivating Rationale

The semiconductor industry began in the 1940s with the invention of the transistor by Bill Schockley, John Bardeen, and Walter Brattain of Bell Labs, followed by the first integrated circuit chip by Jack Kilby of Texas Instruments. The seminal paper by physicist Richard Feynman, "There's Plenty of Room at the Bottom," was presented at an American Physical Society meeting on December 29, 1959, at the California Institute of Technology.[i] This paper brought together notions of micro and nano fabrication, science, and the integration of electronics and mechanical systems; it became the catalyst to kick off the microsystems technology revolution. Since then, the semiconductor industry has maintained continual growth, with the transistor-circuit density doubling every two years ("Moore's Law"), an average compound annual growth rate (CAGR) of 13 percent between 1977 and 2011, and a CAGR of 9 percent in the last ten years.[ii]

As semiconductor chips evolved starting in the 1950s and 60s, the increasingly complex circuits began to be interfaced with sensing units and to incorporate moving mechanical parts. Rather than moving and manipulating only electrons, these systems introduced transduction systems on the micron scale. Not only is mechanical motion sensed and actuated, but MEMS also act as manipulators of light, sound, fluidics, and biological signals.

The MEMS field is continuously growing and evolving. The first device to integrate mechanical and electrical components on a micron scale was the resonant gate transistor,[iii] followed by the monocrystalline silicon diaphragm-based strain-gauge pressure sensor developed by Kulite, also in the 1960s, for aerospace applications.[iv]

During the 1990s, MOEMS (micro-opto-electromechanical systems) devices took center stage in the development sphere both for projection of images and switching optic signals. The most successful MOEMS system is Texas Instruments' (TI) Digital Mirror Device (DMD), part of the Digital Light Processor (DLP) system found in office, cinema, and home-theater projection systems. The most recent systems integrate over eight million moving micro mirrors that are individually addressed and able to switch from off to on position several thousand times per second all on a single integrated MEMS chip. It took an approximately $1 billion investment over ten years for TI to develop the DMD; the first commercial product debuted in 1996. TI acquired approximately $800 million in revenue in 2010 with this single MEMS device,[v] which competes with the ink-jet print head for first place in revenue from year to year.

The most ubiquitous of our modern gadgets is the smart phone, the most advanced having over fifteen MEMS devices, including, but not limited to: accelerometers, a gyroscope, an electronic compass, a pressure sensor, a Bulk Acoustic Wave (BAW) filter, BAW duplexers, Radio Frequency (RF) switches, temperature-compensated crystal oscillators (TCXO oscillators), micro mirror display projection, a complementary metal-oxide semiconductor (CMOS) image sensor, an auto focus actuator, front and rear cameras, proximity sensors, and microdisplays with touch interaction. MEMS cell phone unit shipments composed 45 percent of the total market in 2010 and are expected to grow to 50 percent of the market by 2015. Conversely, automotive market share has dropped from 75 percent of the MEMS devices shipped in 2005 down to 30 percent in 2010, and is expected to be only 20 percent of the market in 2015.[vi] The overall MEMS market continues to grow even in the current economic downturn, as the younger generation considers smart phones a necessity, not a luxury; being socially connected is more important than owning a car or a home.[vii] This also supports the concept of integrating microsystems educational materials to enhance STEM core classes at the high school level.

Students are already enamored of the small devices within the smart phone; meanwhile, biometrics is another engaging topic for students. It is predicted that in the near future, wireless MEMS-based devices "in your home will keep tabs on your medical status every day, as you go about your daily routine."[viii] Biometric sensors and ambient monitoring will become commonplace, and by reducing hospital admissions, they could save up to $6.4 billion annually for the estimated 1.27 million U.S. patients at risk of heart failure. According to Joseph M. Smith, chief medical and science officer of San Diego's West Wireless Health Institute:

We are at an inflection point now, where wireless connectivity, personal cellular devices, pervasive sensing technologies, social networks, and data analytics are mature enough to make wireless medicine a reality. And there is a will as never before to find a way to reduce crippling health care costs. Already, new devices allow diseases like diabetes and chronic heart failure to be closely monitored outside the doctor's office, tools for tracking chronic kidney disease and a variety of lung disorders are sure to follow. Eventually, most health care will occur not during occasional visits to doctors' offices, clinics, or hospitals but continuously, during ordinary activities in people's home.[ix]

MEMS and Microsystems continue to grow at double-digit compounded annual growth rates (CAGR), as predicted by Yole Développement Group, the leading microsystems market research company.[x] In 2010, the global MEMS market was at $8 billion (4 billion units), and it is predicted to grow to $19.5 billion (16 billion units) by 2016. The number of MEMS components in all market segments continues to grow.[xi] For example, the automotive industry, in addition to accelerometers and pressure sensors, now uses mass-flow, angular-rate (rollover), and combined-inertial sensors, as well as microphones.[xii]

The top two MEMS-device manufactures in the world are U.S.-based Texas Instruments and Hewlett Packard, whose combined revenue in 2010 was $1.6 billion. Other U.S.-based companies in the top thirty include: Freescale, Avago, GE Measurement, Analog Devices, Measurement Specialties, Lexmark, Honeywell (a *Southwest Center for Microsystems Education* Industry Advisory Board contributor), FLIR Systems, Triquint, and Goodrich, with a combined impact of another $1.6 billion. Of the top thirty MEMS companies worldwide, which take in a combined $7 billion in sales each year, the United States takes 46 percent of the market share at $3.2 billion.[xiii]

Many MEMS sensors are now being combined into sensor systems, especially in the smart-phone and tablet markets. Since global positioning systems (GPS) do not allow accurate altitude measurements and signals are often distorted in urban environments, systems are coming to market that include inertial, electronic-compass, and pressure (altitude) sensors, so that accurate positioning information can be maintained during periods of signal loss or corruption. This will allow consumers to use their mobile smart phones as navigational tools inside buildings.

As one can see, MEMS typically contain an integrated set of otherwise disparate technologies (e.g., mechanics, fluidics, materials, energy, photonics, biology, etc.) that span the entire spectrum of STEM components. The challenge is to engage and develop an agile, well-educated workforce to support this business growth and provide the full range of needed skills.

MEMS Knowledge and Skills

The *Southwest Center for Microsystems Education* (SCME), along with Central New Mexico Community College (CNM), keeps in continual contact with small-, medium- and large-scale microsystems enterprises. Between 2005 and 2007, SCME completed several projects, including a comprehensive industry survey, a MEMS Technologist Job Profile, and a needs assessment. These projects, along with annual Industry Advisory Board meetings, allow SCME to continuously create and maintain relevant educational materials. SCME is currently engaged in another project to not only identify and map out the microsystems firms in the United States, but also survey a sample to determine the current need for technologists. Using GPS mapping technology, these companies and needs will be overlaid with the locations of community colleges to ascertain where there is sufficient clustering to justify considering implementation of a microsystems program.

Needs Assessment
A regional survey and needs assessment revealed that fifty new microsystems and related jobs were being created in New Mexico annually and that an estimated two hundred additional jobs will need to be filled as our aging workforce retires. Extrapolating this to the nation, it is expected that over two thousand new microsystems technician jobs are created annually.

Job Profiling: Essential Skills

The SCME principal investigator (PI) collaborated with Sandia National Laboratories and David Licht, CNM WorkKeys job profiler, to produce the ACT WorkKeys® job analysis of the MEMS Process Technician job at Sandia National Laboratories (SNL). Eight of the nine WorkKeys® skills are required of a MEMS technician working in a state-of-the-art MEMS fabrication facility: Applied Mathematics, Business Writing, Listening, Locating Information, Observation, Reading for Information, and Teamwork. Over 150 required tasks were observed over several days during the job-shadowing activities. Each task was described, catalogued, and subsequently analyzed. An individual task may require many WorkKeys skills and different levels. A detailed report is available.[xiv]

Industry Input

To better prepare these future technicians to be successful, a comprehensive industry survey was completed.[xv] The results indicated several areas of growth in microsystems. It also indicated that companies with fewer than fifty employees were doing most of the hiring of technicians and that most of the jobs are in research and development and microfabrication, that is, developing and making these small gadgets in clean rooms. The survey results contributed to the selection of instructional topics to be developed.

Specific knowledge areas for this field were also surveyed; the results showed which fabrication methods and basic design principles are most important. Photolithography, etching, and thin film deposition are processes that include principles of chemistry, physics, and some mathematics. Hence, these STEM concepts are included in the educational materials that cover these topics. A basic knowledge of design principles and familiarity with computer-aided-design software is also a plus to be a successful technician. Industry participants ranked knowledge areas, which resulted in the creation of learning modules in the corresponding process technologies. They also ranked STEM knowledge areas; the top five most important subject areas were physics, basic and intermediate mathematics, material science, and chemical safety. Listening, observation, teamwork, computer skills, and time management were the top five most valued "soft skills" and abilities.

The MEMS industry continues to grow, as does the need for technicians. Successful contributors need to have a broad understanding of STEM principles, a global perspective, and good essential skills (listening, observation, teamwork), and must also be multidisciplinary and able to work closely with engineers and scientists.

The Rainbow Wafer Example: Tying STEM to the Real World
A good example of the supporting STEM concepts that are needed to understand microsystems fabrication and design can be seen in the SCME "Etch Overview for Microsystems Learning Module".[xvi] This module describes the materials and fabrication process involved in the deposition of silicon dioxide, a type of glass. The description includes the basic concept of silicon oxidation (similar to the formation of iron oxide, or rust) and the basic chemistry involved in two possible oxidation paths—through direct oxidation of silicon with oxygen gas or through an interaction of the silicon crystal with water vapor. The module also explores the effect on growth rate (micron/hour) as a function of time, temperature, and oxide thickness, and leverages diffusion, mathematics, and graphing concepts.
To create the striped "rainbow wafer," the oxide is etched stepwise by

Figure 11.1. Silicon dioxide coated wafer that has been selectively etched to produce varying thinesses of remaining oxide (above)

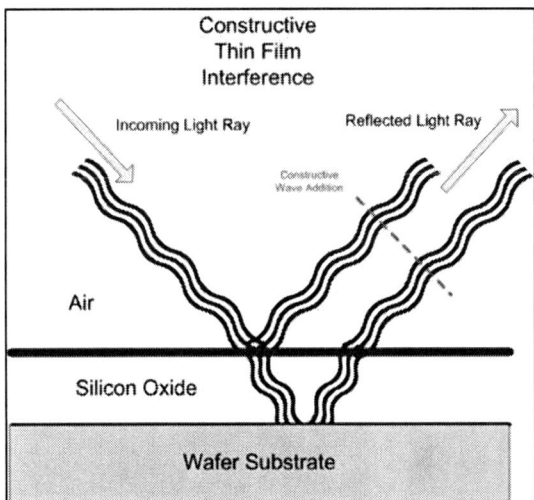

Figure 11.2. Thin film interference effect

sequentially lowering the wafer in a hydrofluoric acid (HF) solution. Each stripe is etched for an increasing amount of time. The resulting thin silicon oxide film stripes each appear as a different color due to the thin film interference effects of light reflecting off the top surface of the silicon crystal substrate and the top surface of the silicon oxide glass. The color of each stripe is cross-referenced with a color chart to determine the actual silicon oxide thickness.

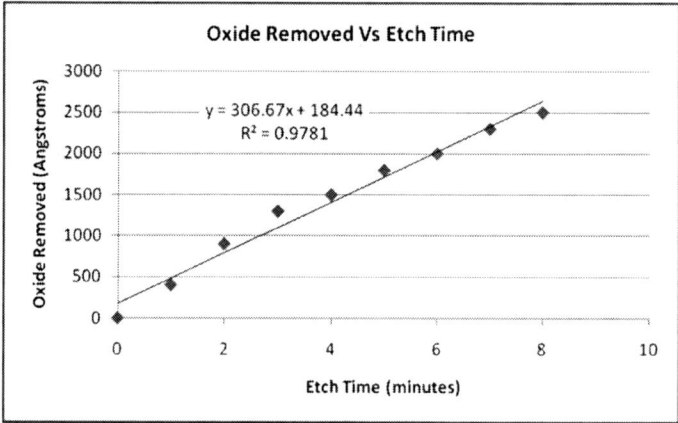

Figure 11.3 Straight line fit to determine etch rate—math applied to a real world high-tech problem.

Using these data, the students plot oxide thickness versus etch time and determine the etch rate by applying a straight-line fit to determine the slope. This is then used to determine the time required to remove a given amount of oxide. In summary, the students use STEM concepts in the application of thin deposition, etching, and thin film measurement as used in microsystems fabrication to solve a typical problem encountered by a MEMS technician. When taught in secondary schools, this learning module ties STEM learning to a future job function, giving students an answer to the age-old questions "When will I ever use that?" and "Why I should care?"

Figure 11.4. Preparing a furnace tube for a silicon oxide growth process

Learning Modules and Hands-On Kits

The SCME, based on industry input and the industrial experience of the PI (twelve years in semiconductor fabrication as a process engineer, equipment engineer, and yield and metrology manager), decided early on to develop a series of learning modules around the fabrication of a simple MEMS device. SCME used the device of choice, the piezoresistive bridge pressure sensor, to teach materials properties, basic electronics, and bulk and surface micromachining, along with the corresponding STEM concepts and essential skills. A pressure-sensor device was developed at the University of New Mexico's Manufacturing Training and Technology Center, and subsequent educational materials were vetted at several five-day pressure-sensor clean-room professional-development workshops. The workshop materials were further developed, used and tested in several MEMS courses conducted at Central New Mexico Community College.

After hosting numerous workshops and giving many presentations, all the while acquiring participant feedback for continuous improvement, SCME saw a need for educators to be able to bring clean-room experiences back to the classroom. Hence, SCME developed eleven kits to enable this kind of practical classroom experience, and several more are currently being planned). Below is a complete listing as of 2011.

Each of the following learning modules (Shareable Content Object [SCO] suites) includes guides for instructors and students. Each module has at least one of each type of SCO, beginning with primary knowledge and followed by activity and assessment SCOs. For many of the learning modules, SCME has included module maps and STEM standards in the instructor guides to assist educators in implementation.

Introductory Topic Modules (SCOs)
- MEMS History (3)
- Units of Weights and Measures (4)
- A Comparison of Scale—Macro, Micro, and Nano (5)
- Introduction to Transducers, Sensors, and Actuators (3)
- Wheatstone Bridge Overview (4)
- Cantilevers I & II (6)

Application Modules (SCOs)
- MEMS Applications (3)
- BioMEMS Applications (5)
- Microcantilever Learning Module (7)
- Micropump Learning Module (4)

Fabrication Modules (SCOs)
- MEMS Micromachining Overview (4)
- MEMS: Making Micro Machines (5)
- Manufacturing Technology Training Center (MTTC)—Pressure Sensor Suite (6)
- Crystallography for Microsystems (5)
- Photolithography Overview (3)
- Etch Overview (4)
- Deposition Overview for Microsystems (3)
- Oxidation Overview for Microsystems (1)
- LIGA (4)
- MEMS Innovators Wanted (8)

Safety Modules – (SCOs)
- Hazardous Materials (5)
- Chemical Lab Safety Rules (3)
- Material Safety Data Sheets (6)
- Chemical Labels/NFPA (5)
- Personal Protective Equip.—PPE (3)

BioMEMS Modules (SCOs)

- What are BioMEMS? (3)
- Mapping Biological Concepts:
  - DNA Overview (5)
  - DNA to Protein (3)
  - Cells—The Building Block of Life (3)
- Biomolecular Apps. of BioMEMS (6)
- BioMEMS Diagnostics Overview (3)
- BioMEMS Therapeutic Overview (3)
- Clinical Laboratory Techniques & MEMS (3)
- Regulations of BioMEMS (4)
- MEMS for Environmental & Bioterrorism (3)
- DNA Microarrays (6)

Hands-on Kits

1. Anisotropic Etch
2. Crystallography
3. Dynamic Cantilever
4. Lift Off
5. MEMS Innovators
6. MEMS: Making Micro Machines DVD
7. Pressure Sensor Model
8. Pressure Sensor Process
9. Rainbow Wafer
10. LIGA Micromachining
11. Gene Chip Model

The learning modules include original graphics and animations for educators to use in their classrooms and are all available through the SCME website for free downloading. Printed copies and kits are provided through SCME website requests and are facilitated by the nonprofit Center for Hands-on Learning, located in Rio Rancho, New Mexico. SCME has also started a webinar series that promises to reach a larger number of educators. These webinars are available for secondary and postsecondary educators and their students to view at any time. These webinars are facilitated by SCME's partner, Arizona's Maricopa Advanced Technological Education Center (MATEC).

All the kits and most of the learning modules are now being used by high school STEM teachers. Over 50 percent of the professional development participants are high school educators. Teachers really like to use the hands-on kits. For example, both the Anisotropic Etch and Lift Off kits are commonly used in chemistry classes to reinforce etching (dissolution) of materials, and the associated chemical reactions are applied to a real life examples—for example, the making of the tiny devices found in students' cell phones. The safety modules on Materials Safety Data Sheets, NFPA symbols, Personal Perfective Equipment, and the like are also used in many chemistry classes. As another example, the Cantilever Kit allows a physics or engineering teacher to present a variety of STEM topics:

- Materials science (Young's modulus)
- Data acquisition and analysis
- Graphing nonlinear relationships
- Classic "mass on spring" problem
- Curve fitting real data to theory
- Writing a lab report
- Video analysis
- Resonance frequency
- Teamwork
- Natural frequency of systems
- Position, velocity, and acceleration (applied calculus)
- Extrapolation and application to the micro scale
- The overlap of biology and microtechnology is one of the fastest growing segments in the microsystems industry, and as mentioned earlier, it's another engaging topic for students.

It is fortuitous that the most popular science class is biology, hence the popularity of the series of modules found under the BioMEMS heading. The GeneChip kit is also a popular hands-on activity used in both high school and college classes. This micro device allows one to test for hundreds of gene sequences on a single centimeter-square microchip. MEMS devices are on a path to dominate not only the disease-detection and sensing industries, but also the replacement and augmentation of biological systems (examples include cochlear implants, stents, pacemakers, glucose-monitoring devices, insulin micropumps, artificial retinas, neural probes, and camera pills).

Professional Development and Impact

Preparing secondary and postsecondary educators is a major goal for SCME. SCME seeks to advance the implementation of MEMS-based curricula as part of technician programs and STEM courses in both high school and postsecondary settings. The modular design of the learning modules allows educators to leverage specific elements and integrate those elements into their curriculum as it makes sense for their institution and course. For example, chemistry teachers might use the Anisotropic Etching module in their classes to show students the application of crystallography principles; the (100) and (111) crystal plane orientations have different respective etch rates in KOH (potassium hydroxide). These different etch rates allow one to form specific microsystem structures such as pressure sensor cavities or microfluidic channels. The students actually observe the etching of the back side of a pressure sensor chip in a beaker of heated sodium hydroxide solution (drain cleaner) in their chemistry classroom. This not only brings together principles of chemistry, safety, and materials science, but also introduces students to the idea of making a career out of producing the tiny gadgets that are found in a plethora of modern products. It also gives students an incentive to learn chemistry and other subjects as steps toward a future career.

The pinnacle of workshop training for high school and college educators consists of a four- to five-day pressure-sensor clean-room workshop. Several of these workshops have been conducted in recent years at the University of New Mexico's Manufacturing Training and Technology Center (MTTC). During the workshops, six to twelve educators spend 50 percent of their time in the clean-room making pressure sensors; the other 50 percent of the time, they are being trained in the use of the associated kits so that they can bring their clean-room experience back to the classroom. A complete list of kits the educators use during this weeklong workshop includes:

1. Pressure Sensor Macro Model. Students make a working model of a strain-based membrane pressure sensor. Wheatstone bridge circuit principles are explained and demonstrated.
2. Crystallography. Students learn the basics of crystal, polycrystal, and amorphous structure, as well as Miller Indices

as applied to crystal plane. Activities include making a model that shows the planes and breaking (111) and (100) wafers.

3. Anisotropic Etch. This kit explains and demonstrates back-side etching to create a chamber below a silicon nitride membrane.

4. Lift Off. This is the final step of a photolithography, metal deposition, and liftoff process, whereby a gold circuit pattern (Wheatstone bridge) is transferred to the surface of a silicon nitride membrane. Kit materials describe metal deposition and the coat/expose/develop photolithography process.

5. Pressure Sensor Process. This kit includes two sets of ten pressure-sensor chips taken from ten process steps. Materials explain the details of the process, and students have an opportunity to observe the construction of a pressure sensor from bare silicon to a completed, working chip.

6. Rainbow Wafer. This kit explores the thin film interference effect used to determine the thickness of silicon oxide thin films. This optics principal is explained and applied to determining the etch rate of silicon dioxide.

7. MEMS: Making Micro Machines DVD. This kit includes a DVD and three activities to aide in the understanding of how MEMS DLP and ink-jet printheads are made as well as how motion sensors are designed to meet customer requirements.

8. MEMS Innovators. This capstone activity, designed for more advanced workshops and students, brings together design and fabrication principles to solve a real-world problem. Students design a MEMS device, fabricate a model, and present it to the class. This activity is typically done during the five-day version of the pressure-sensor clean-room workshop.

In August 2011 and January 2012, SCME completed technology-transfer events for the North Dakota State College of Science (NDSCS) Center for Nanoscience Training Technology (CNTT) and for the University of South Florida (USF) Nano Research and Education Center (NREC). These two clean-room "hub" facilities can now process and produce the SCME MEMS pressure sensors, and staff members have been trained to conduct the associated kit workshops.

Educators are also trained in these and other kits during half-day and one-day workshops at conferences and other events:

1. Dynamic Cantilever. These structures are the basis of a wide variety of MEMS devices, including the resonant gate transistor, atomic-force microscope (AFM), chemical-sensor arrays, hard-drive heads, and cantilever-array storage devices (such as IBM's Millipede). The physics principles of the resonance are demonstrated as applied to the cantilever (harmonic oscillator). Students learn data-acquisition techniques that use video acquisition and analysis applied to determining the natural frequency as a function of material properties (bulk modulus), cantilever dimensions (thickness, width, and length), and mass added. Students learn data acquisition, analysis, graphing, lab-report writing, and predicting through theory at what frequency a microcantilever of given dimensions made of silicon will resonate.

2. GeneChip. This model of a DNA gene chip process uses lithography principles coupled with the deposition of DNA components. These GeneChips are utilized in diagnostic and drug discovery applications. This is a favorite of biology educators and provides an entry point for this traditional STEM subject to bring MEMS into the curriculum.

3. Pressure Sensor Model. Students build a working pressure-sensor model in a classroom environment. This device allows the STEM educator to give the students an opportunity to explore through hands-on experimentation how to fabricate, test, calibrate, and evaluate a real working device. The simple Wheatstone bridge circuit is made up of finely ground graphite (pencil lead) mixed with rubber cement and applied to the surface of a balloon stretched over a paint can. An input voltage (battery) is applied, and the output is measured with a voltmeter. The output is related to the force (pressure) applied. Basic mathematics is introduced relating the resistance of an element to its composition, length, and cross-sectional area. As with the cantilever kit, students reinforce their mathematic, graphing, problem-solving, and analysis skills.

Central New Mexico Community College Advanced Systems Technology Program: A Process of Evolution

Central New Mexico Community College (formerly known as Albuquerque Technical Vocational Institute) initiated a MEMS technician program under the Manufacturing Technology AAS degree program in 2004. The first course developed was the Introduction to MEMS three-credit-hour lecture/lab course. The second course to be implemented was the MEMS Fabrication course, which was a five-credit-hour lecture/lab course; this was followed closely by two design courses, MEMS Design I and II (which were each three credit-hours). These new courses replaced several Semiconductor Manufacturing Technology (SMT) courses because Philips Semiconductors shut down (Albuquerque), and Intel began reducing their SMT workforce due to continued automation. The microsystems or MEMS program evolved over time as a result of local CNM-industry input and SCME national industry surveys as well as feedback from SCME's industry advisory board. This original program was built on a traditional electronics foundation with MEMS and manufacturing technology courses.

Within the past year, SCME and CNM have worked collaboratively to develop the Advanced Systems Technology program, which more appropriately addresses student, institution, and industry needs. New Mexico is a hotbed of technology companies that have emerged from companies ranging from small start-ups, such as Life BioSciences and HT Micro, to medium-sized organizations, such as Emcore and CVI Melles Griot, to the very large-scale enterprises of Intel and Sandia National Laboratories. These organizations conduct research and manufacturing on a wide variety of micro- and nano-related products; some also provide services to user organizations. As a result, this variety of industry stakeholders prefers technicians with a high level of hands-on skills, as well as familiarity with basic electronics, lab-view, manufacturing, and microsystems-design concepts. They want technicians with fabrication knowledge and skills, including familiarity with clean-room processes and protocols, photonics (optics) concepts and their hands-on applications, and essential (soft) skills. This is in line with the Defense Advanced Research Projects Agency's (DARPA's) Microsystems Technology

Office 1999 projection for integrating electronics, photonics and MEMS (EPM); DARPA aimed to accomplish true "chip-scale" integration of these core technologies along with computer-aided design (CAD)[xvii] and simulation tools.[xviii]

The following curriculum framework describes the latest evolution of the MEMS program at Central New Mexico Community College. It can be a model for other technology-based two-year programs whose directors wish to develop them to meet the needs of emerging micro- and nano-based industries.

Career Pathways: High School to Industry
The Introduction to MEMS course is offered as a dual-credit/dual-enrollment course at CNM to area high schools. Only one teacher currently teaches Introduction to MEMS. However, high school students and teachers do take the course at CNM. The PI of the SCME has developed the intro course into a hybrid course that takes place partly online and partly face-to-face. The face-to-face component is offered in the evenings or on Saturday mornings to better accommodate nontraditional student, teacher, and high-school-student schedules. Recently, as part of the new Advanced Systems Technologies program, all the MEMS courses have been formally split into lecture/lab sequences so that the lecture portions can be offered online through distance learning and the face-to-face labs can be offered in a variety of ways. Currently, the labs meet weekly; however, in the future, the labs could be offered in one- or two-week intensive sessions so that students in remote locations could travel to the University of New Mexico's MTTC or one of the hubs (North Dakota State College of Science's CNTT, University of South Florida's NREC) after taking the lecture portion either at their school or online. This is along the same lines as how SCME currently offers the Pressure Sensor Clean-Room Workshop to educators at the three hubs.

Many teachers incorporate MEMS Learning Modules into their high school STEM classes. This may not immediately result in a students following a MEMS career pathway, but it does provide them with an option that they would not have considered before. SCME plans to continue working with local high schools to formalize the integration of MEMS kits and modules as anchoring activities in STEM classes.

Albuquerque Public School System has agreed to cross-train 72 STEM teachers per year over four years to bring these modules and career-education units to the classroom. This has the potential to affect more than ten thousand students and 160,000 student hours. The following are planned for APS high school:

- Biology: MEMS DNA Chip Array, BioMEMS Applications
- Physics: Dynamic Cantilever Kit, Pressure Sensor Kit, MEMS Applications
- Chemistry: Anisotropic Etch, Lift-off
- General Science: MEMS Applications, Crystallography Kit (Origami, Breaking wafers, Miller indices)
- And for middle and high school science courses: Safety Suite of Learning Modules.

SCME has also begun to train educators at the ASK (Alternative School of Knowledge) Academy, a STEM-based, pre-engineering charter high school. The plan is to have this school also integrate MEMS modules into the classroom and offer Introduction to MEMS to their junior-level students, to be followed by the MEMS Design course. These will be taught by ASK teachers and articulated to CNM. It is clear that many of the ASK Academy students will not have the desire or necessary skills to be successful at a University Engineering program; hence, the technician route is a good path for these students. Many CNM technician graduates have continued in an engineering program after being successful in the workplace and maturing to a point where they could be successful in a calculus-based engineering program.

CNM MEMS-program graduates have placement rate of over 95 percent; they have been employed by companies such as Intel, Sandia National Laboratories, Emcore, TPL, Cardinal Health, Avago, and Thunder Scientific.

Concerns

The most difficult issue for these emerging technology programs is producing enough technicians to support the ever-increasing needs of industry. Recruiting students to pursue these two-year careers to adequately fill the courses is an ongoing challenge. There are many reasons for this. The term "manufacturing" has had a "dated" connotation; it has only recently been recognized as the process of producing modern, high-tech devices that are in demand. For the last twenty years, the public perception was that all manufacturing work was being sent offshored and that the United States would become the global leaders of financial services. We have seen the result of that. "Manufacturing" conjures up visions of a dirty, sweaty industry. And when people do think about high-tech manufacturing, they often remember the large number of layoffs in the semiconductor industry. The general public does not recognize the terms "MEMS" and "microsystems," so education about these technologies is also needed, and it needs to start in middle school.

The key is *jobs*. Over and over again, we hear from high-tech industries that they cannot find enough qualified technicians, engineers, and scientists. There are approximately four to five technicians needed for every engineer in medium- to large-scale high-tech industries. It is not common for teachers and counselors in high schools to encourage students to pursue a technician degree, even though most high-tech technicians earn more than liberal-arts bachelors-degree recipients.[xix] Only one-third of students who start a STEM degree program (for example, engineering) actually acquire a degree in a STEM field; the attrition increases to 80 percent for students from underrepresented populations.[xx] These tech-curious and tech-savvy students leave their engineering major for a variety of reasons; but many do not know they can have a successful and well-paid career as a technician. There are many reasons for the high dropout rate in engineering; the subject matter is very difficult, and most first-year college students don't have the necessary mathematics foundation of high school calculus. Hence, they are underprepared in many STEM disciplines and have to catch up. Most students take five to six years to complete the BS in engineering. Many first-year college students don't have the maturity and study skills needed for such a rigorous program. Cost can also be an issue, as many state programs only sponsor students for four years.

A technician graduate can go to work two years after high school and make more than $40,000 per year—more than a high school STEM teacher with a four-year degree. Some highly skilled and experienced technicians earn more than engineers, over $100,000 per year. Technicians keep the high-tech manufacturing facilities running by monitoring processes; implementing corrective action plans; and maintaining, installing, and repairing equipment valued in the $1 million to $10 million range.

There is a solution: encourage more students to pursue technician degrees, especially those students who prefer hands-on activities and have an inclination to take things apart. Use hands-on applied topics centered on high-tech jobs to give these students high school experiences that will cause them to pay attention and learn STEM concepts. Many technician graduates begin their careers at high-tech companies and government labs. Once on the job, they continue to improve their critical thinking skills and gain maturity, which allows them to succeed in engineering school if they desire to pursue that path later on. Companies often support this pursuit. Community colleges need to partner with local high schools to increase this pipeline of technicians who will, in turn, experience rewarding careers.

---

[i] Richard P. Feynman, "There's Plenty of Room at the Bottom: An Invitation to Enter a New World of Physics," *Journal of Microelectromechanical Systems* 1.1 (1992): 60–66, online at http://www.zyvex.com/nanotech/feynman.html (accessed March 21, 2012).
[ii] Semiconductor Industry Association, "Historical Billing Reports: February 6, 2012," http://www.sia-online.org/industry-statistics/historical-billing-reports/ (accessed March 21, 2012).
[iii] Harvey Nathanson, "The Resonant Gate Transistor," IEEE Transactions on Electron Devices ED-14.3 (1967): 117-133.
[iv] Kulite, "Kulite's Roots," http://www.kulite.com/docs/history/1959-2009.pdf (accessed March 31, 2012).
[v] Jérémie Bouchaud, "DLP Revival Returns Texas Instruments to MEMS Market Leadership in 2010," April 5, 2011, http://www.isuppli.com/mems-and-sensors/news/pages/dlp-revival-returns-texas-instruments-to-mems-market-leadership-in-2010.aspx (accessed March 31, 2012).

vi SEMICO Research Corporation, "MEMS Growth: Changing Markets, Changing Applications," March 22, 2011, http://www.semico.com/press/press.asp?id=277 (accessed March 31, 2012).
vii Nielson Company, "Mobile Youth Around the World," December, 2010, http://www.nielsen.com/us/en/insights/reports-downloads/2010/mobile-youth-around-the-world.html (accessed March 31, 2012).
viii Bob Charette, "Digital Technologies Will Soon Deliver Daily Diagnoses," *IEEE Spectrum Computerwise*, October 5, 2011, http://newsmanager.commpartners.com/ieeecw/issues/2011-10-05-email.html (accessed October 10, 2011).
ix Joseph M. Smith, "Wireless Health Care," *IEEE Spectrum Inside Technology*, October 2011, http://spectrum.ieee.org/biomedical/devices/wireless-health-care (accessed October 10, 2011).
x Yole Développement, 2011, http://www.yole.fr/home.aspx (accessed June 7, 2012).
xi Jean-Christophe Eloy, "MEMS Industry Evolution: From Devices to Function" (Semicon West presentation, 2010), http://semiconwest.org/sites/semiconwest.org/files/JC%20Eloy.Yole_.pdf (accessed June 7, 2012).
xii Gary O'Brien, *New MEMS Devices Aimed at Emerging Consumer and Automotive Applications* (SEMICON West presentation, July 12, 2011), http://semiconwest.org/sites/semiconwest.org/files/Gary%20O%27Brien_Bosch.pdf (accessed June 7, 2012).
xiii Ibid.
xiv David Licht, "WorkKeys Job Profile Report: Microelectromechanical Systems, Sandia NationalLaboratory," June 6, 2006, http://scme-nm.net/scme_2009/files/MEMS_profile.pdf (accessed March 21, 2012).
xv Matthias W. Pleil, "Knowledge and Skills Needed by MEMS Technologists as Ascertained by Industry Survey and Job Profiling," ATE PI Conference, October, 10, 2007, http://scme-nm.net/scme_2009/files/knowledgeskills_memstech.pdf.
xvi Southwest Center for Microsystems Education, "Etch Overview for Microsystems Learning Module," June 13, 2011, http://www.scme-nm.org/scme_2009/index.php?option=com_docman&task=cat_view&gid=95&Itemid=53 (accessed March 31, 2012).
xvii William Tang, "MEMS Programs at DARPA," ME219 Lectures at University of California, Berkeley, January 1, 2011, www.me.berkeley.edu/ME219/Lectures_2011/L43ME219.pdf (accessed March 31, 2012).

[xviii] Noel MacDonald, *Microsystems Technology Office* (DARPA Tech presentation, 1999),
http://archive.darpa.mil/darpatech99/Presentations/mtopdf/mtooverview.pdf (accessed June 7, 2012).
[xix] Jason Koebler, "STEM Associate's Degree Pays Better Than Liberal Arts Bachelor's," *U. S. News and World Report*, February 4, 2012,
http://www.usnews.com/news/blogs/stem-education/2012/02/24/stem-associates-degree-pays-better-than-liberal-arts-bachelors (accessed June 7, 2012).
[xx] Higher Education Research Institute (HERI), "Degrees of Success: Bachelor's Degree Completion Rates among Initial STEM Majors," *HERI Research Brief* (January 2010),
http://www.heri.ucla.edu/nih/downloads/2010%20-%20Hurtado,%20Eagan,%20Chang%20-%20Degrees%20of%20Success.pdf (accessed June 7, 2012).

Secondary/Postsecondary Programs of Study for Materials Technicians

Mel Cossette and Frank Z. Cox
National Resource Center for Materials Technology Education
(MatEd)

Introduction

MatEd (the National Resource Center for Materials Technology Education, a National Science Foundation Advanced Technological Education funded center) builds upon numerous public and private initiatives, including other NSF-funded centers and projects, through strong partnerships with industry and higher- and secondary-education professional groups.

MatEd's key goals are to: a) broaden access to resources for materials technology education faculty and other professionals (www.materialseducation.org); b) improve and broaden the Center's resource collection, which includes high-quality, peer-reviewed technician-education modules in both traditional and new advanced materials technologies; c) broaden dissemination strategies through an enhanced website, increased partnerships, stronger regional and national support systems; and d) implement a long-term program of sustainability.

With rapid changes and new developments in nanotechnology, nanoscience, corrosion, and green and composite technologies, as well as the more traditional areas of metals, plastics, and composites, it is imperative for industry to have qualified and educated technicians to keep the United States globally competitive in materials and advanced manufacturing.

Future of Materials Technology Education

Materials engineering technologies are the basis of many primary manufacturing processes, and the properties of the materials used in structures and devices can determine the lifetime and behavior of those systems. The properties of these materials depend on the materials' structure and processing, which create variations that can directly affect a product's function and life cycle. Materials involved are plastics, metals, composites, wood, and ceramics, and these materials have applications in structures, electronics, and biological and medical systems, including recent advances in the use of nanotechnology.

Programs often find it difficult to keep up with changing technological applications. According to Kevin Gulliver, president of Nida Corporation, technicians often graduate from entry-level technical training programs with the wrong skill sets to properly meet the challenges of today's complex technical systems.[i] Gulliver focuses on the need for active learning using real problems and real systems; this is similar to placing an emphasis on hands-on experiential learning in both the K–12 system and in higher education. According to Conrad Ball, Chief Engineer at Boeing Military Aircraft, "Four-year degrees are important, but I want you to know that two-year technical people are also very important. I struggle every day to find skilled people." Ball also stated that "for every engineer we hire, we estimate it affects up to nine additional positions [technicians] within our company."[ii]

The potential future lack of technicians trained in composites technology prompted The Boeing Company, for example, to develop a joint program with MatEd and nearby Edmonds Community College to provide a hands-on internship experience and incentives to enhance the training of technicians in Edmonds' Materials Science program. Fortunately, materials technologies are excellent subjects for introducing hands-on learning at all educational levels, and a wide variety of experiential instructional resources are available. National Educators Workshop 2012 evaluation feedback regarding hands-on activities included comments such as, "amazingly useful," "essential for information to help start a materials program at my school," and "one of the most relevant workshop for hands-on information."[iii]

For our country to maintain its position in the world economy and continue to be on the cutting edge, increasing the quantity and quality of materials engineering technology education is important due to several factors, including:

- The increased complexity of materials and engineering systems;
- The economic effects of errors in materials processing that lead to materials and systems failures;
- The economic effects of corrosion, wear, and related materials-degradation processes on equipment and infrastructure;
- The need for enhanced system performance while minimizing weight, cost, and time to market; and
- The emerging requirements for environmentally friendly systems and product recycling.

This chapter discusses some of the current efforts at enhancing materials technology education and identifies specific areas for future focus in STEM programs.

Materials Engineering Technology Education

Materials technology education provides both technicians and engineers with the needed understanding of the relationship between a material's properties, its structure, and the proper means for processing the materials to arrive at the needed strength, ductility, hardness, or surface condition. While a typical process is often prescribed by a set of written directions or standards, variations in understanding and application of these rules can lead to material (and system) failure. "The furnace is not hot enough, but we went ahead anyway" or "the surface did not look quite right, but we thought that our coating would solve the problem" are two of many similar comments heard during failure-analysis investigations. System-design errors may lead to other problems, such as accelerated corrosion, fatigue failures, or surface irregularities.

For example, manufacturing and engineering technicians are required to manufacture parts using a variety of techniques that can change the properties of the materials involved. Often they are in charge of a process for joining parts (gluing, welding, brazing); coating parts for corrosion- or wear-resistance (plasma spraying, electroplating); or adjusting color or appearance (anodizing, patterning). Other processes technicians use to manufacture parts include injection molding, layering, casting, rolling, drawing, forging, and heat-treating.

Slight changes of conditions during any of these processes can modify the resulting material's properties. Electronics manufacturing uses processes such as plating, etching, and soldering, in which outcomes can also be highly process dependent. Semiconductor and microsystems technologies require yet a different set of skills in etching, deposition, and testing.

Future technician education programs need to integrate both traditional materials selection and processing and more advanced methodologies. Programs should be tailored to industry needs, which often change as materials technologies evolve.

The basic question here is: *What should an engineering/materials technologist know about materials, now and into the future?*

To determine at least part of the answer to this question, MatEd conducted surveys and focus groups, analyzed the data, and developed core competencies for general technicians and technicians working in the materials areas. Building on this work and additional research, MatEd developed core competencies for marine technology and nanotechnology. These studies are available at http://www.materialseducation.org/educators/competencies. While these studies focused on technicians, their findings can also apply to engineers.

In terms of materials technology education, these studies focused on the need for education related to materials structure and microstructure and the relationship of structure to materials properties. This also extends to materials-processing technologies and the effects of processing on properties and thus on system performance. One key point these studies identified is the importance of careful attention to processing parameters such as temperature, time, speed, and other variables that may affect the result and thus how the part performs within the resulting product or system.

In terms of the future of materials technology education, MatEd's competency studies suggest that *all* engineers and technicians need to learn the basics of materials technology, and especially the important relationships between structure, processing, properties, and system performance. In general, it's essential for engineering and materials technology programs to provide deeper exposure to these concepts.

One means of meeting this critical need is providing modules that focus on specific core competencies; these modules can then be inserted into current programs and courses. A wide variety of sources for such modules are provided as resources on the Internet.[iv] Another means is providing opportunities for instructors to broaden their own educational background so that they understand how critical it is to include materials issues in technology education. The first step to meeting this need, though, is to raise awareness of the importance of materials technology education across the entire spectrum of technology.

Issues in Materials Applications

Several subjects offer specific opportunities for enhanced training for technologists in science, technology, engineering, and mathematics (STEM). Most of these STEM subjects are treated broadly in the materials and manufacturing literature.

1. Materials Selection in Design. In a design process, it is critical that potential materials be identified early in the concept phase. Technologists must choose materials whose properties are consistent with the potential design. They must also consider available means of processing the material for the specific application; long-term use, wear, and exposure conditions; and resources for recycling and disposal at the end of the product's lifecycle. For today's complex systems, these tasks require materials knowledge beyond that found in specification tables and handbooks.

2. Environmental Degradation and Corrosion. The National Association of Corrosion Engineers has estimated that the total annual direct cost of corrosion in the United States is $276 billion, over 3 percent of the gross domestic product (GDP).[v] Just for military equipment, the infrastructure costs due to corrosion in 2011 were approximately $22.5 billion.[vi] Corrosion is an electrochemical process aided by differences in electrochemical potential between different metals and other conductors (such as graphite composites) in a system. Unfortunately, education in corrosion is not widespread among

engineers or technicians, with the exception of those working specifically in corrosive environments such as marine technology, aerospace, and biomaterials and implants. This area is especially important in design education.

3. Nanotechnology. This is one of the developing areas of materials technology that is evolving, quickly increasing the number of materials and applications. Here, the influence of material processing parameters can be even more important than for more traditional materials. Since the material properties may depend on the distribution of nanoparticles, for example, errors in processing may influence properties even more. Control of processing requires specific equipment and training in this technology, which has applications from the development of alloys with enhanced properties and the development of microsystems to the development of nanoscale drug-delivery systems and other new medical technologies.

These are three examples of rapid technological advancements that will require in-depth education for future engineers and technicians. Several obstacles stand in the way of broader and deeper training in these subject areas, including:

- Lack of student awareness and knowledge in these subjects and a lack of student interest due to limited traditional teaching techniques;
- Lack of knowledge and understanding of new technologies on the part of existing faculty and trainers; and,
- Lack of funding for expanding or enhancing academic programs.

We can encourage and stimulate student interest and learning through more applied STEM curricula in K–12 education. Several studies have referenced this theory, and some progress has been made through the use of everyday science applications in the classroom.[vii] High school technology courses are making a comeback in some school districts, and these courses provide students with a means of assessing the field. Overall, using theory to solve real-life challenges enhances student interest in and knowledge about the materials technology area; this is true in both high school science classes and technical college introductory courses.

Industry can also play a strong role in materials technology education, since the need for technological training and education will only continue to increase, especially as our aging workforce retires. One solution is to provide internships for students to introduce them to the field, to enhance their motivation to continue on in materials technology,[viii] and to improve their understanding of available career pathways.[ix] In the Boeing Company's case, this involves a community college program in which students can choose between a certificate or degree program of study. Further training and education comes through exposure to different materials and testing labs, on-the-job training, specific courses, and experience. As is often the case, students who successfully complete the internship are more employable, at higher wages. The Boeing Company encourages student completion of the Associate Applied Science-Transfer Degree and will pay tuition for employees. Boeing also has tuition assistance for bachelor's degrees in engineering and technology. An advantage of the Materials Science Associate Transfer Degree is that the degree articulates with programs such as Oregon Institute of Technology's BS programs in Mechanical Engineering Technology and Manufacturing Engineering Technology, programs that are delivered on site for Boeing employees. Articulation with Central Washington University, Western Washington University, or Embry-Riddle University Aeronautical University is also an option.

The chart below shows a materials science career pathway, illustrating entry points, descriptions of certificates and degrees, options to move into bachelor's programs, and earning potential. The chart emphasizes skill and wage progression and hands-on and project-based learning. According to the U.S. Department of Labor, technicians earn from $13.00 per hour for basic assembly and manufacturing jobs to $27.89 per hour for engineering, manufacturing production, aerospace, and nondestructive testing jobs.

Entry Point	Credential	Wage and Occupation Information
Quality and Manufacturing Certificate of Completion (15 credits) MST 110—Manufacturing Basics MST 120—Introduction to Metrology MST 130—Quality Assurance Tools	Certificate of Completion	Assemblers and Fabricators $13/hr Fiberglass Laminators and Fabricators $13.75/hr Dental Laboratory Technicians $16.90/hr Quality Assurance: Inspectors, Testers, Sorters, Samplers, and Weighers $15.88/hr
Materials Science Certificate of Completion (15 credits) MST 175—Introduction to Materials Science MST 180—Polymer Technology MST 200-- Introduction to Composites These courses apply to both the ATA and AAS-T Degrees	Certificate of Completion	Composite Assembly Technicians Composite Lay-up Technicians Composite Quality and Testing Technician Composite Manufacturing Technician Approximate Wage Range $14.00–$18.00/hr
Manufacturing and Materials Associate of Technical Arts Degree (90 credits) This two-year degree gives graduates a basic understanding of manufacturing procedures and processes. Graduates will be competitive for entry-level jobs in manufacturing and aerospace industries.	ATA Degree	Medical Appliance Technicians $17.14/hr Aircraft Structure, Surfaces, Rigging, and Systems Assemblers $21.55/hr Engineering Technicians $27.89/hr
Materials Science Technology Associate of Applied Science–T Degree (90 credits) This two-year degree prepares graduates to enter the industry or transfer to a four-year college or university. This program overlaps with the ATA program: the second year of the ATA program is also the first year of the AAS–T program.	AAS–T Degree	Aerospace Engineering and Operations Technicians $27.93/hr Medical Equipment Repairers $21.39/hr Manufacturing Production Technicians $27.89/hr Nondestructive Testing Specialists $27.89/hr

Figure 12.1. Composites Technology/Materials Science Career Pathways. Adapted from an original chart published by Edmonds Community College.

Conclusion

MatEd is committed to developing adaptable resources that enable courses and programs to be enhanced with up-to-date information about materials and processes technology. It is especially important to stay up to date with emerging technologies, such as those involved in structural manufacturing and electronics, system design, and processing for environmental safety. MatEd initiatives build upon existing materials science curricula to enable faculty to improve technician education and meet the changing needs of industry. By disseminating information about emerging topics, MatEd continues to meet the current and future educational needs of engineering and manufacturing technicians throughout the nation. A set of fully cataloged, cross-referenced, and peer-reviewed modules, demonstrations, and laboratory exercises are available on the MatEd website, at http://www.materialseducation.org.

---

[i] Kevin Gulliver, "Systems Analysis: Training Technicians for Today's Technical Challenges" (paper presented at National Educators Workshop, Greensboro NC, October 16-18, 2011).
[ii] Conrad Ball, "Industry and Emerging Technological Trends" (presentation at *STEM Summit: Integrating STEM onto Today's Classroom to Develop Tomorrow's Leaders*, February 9-10, 2012).
[iii] National Resource Center for Materials Technology Education (MatEd), Survey of Participants, *National Educators Workshop: Converging Technologies and Disciplines*, October 16-18, 2011.
[iv] Materials Technology Modules and many other Instructional Resources are available on the National Resource Center for Materials Technology Education website at http://www.materialseducation.org.
[v] Gerhardus H. Koch, Michiel P. H. Brongers, Neil G. Thompson, Y. Paul Virmani, and J. J. Payer, J., *Corrosion Costs and Preventative Strategies in the United States* (NACE International, Publication No. FHWA-RD-01-146, 2001), http://www.nace.org/uploadedFiles/Publications/ccsupp.pdf (accessed May 17, 2012).
[vi] Ibid.
[vii] MatEd Survey of Participants, *National Educators Workshop*, October 16-18, 2011.
[viii] George Parker and Imelda Cossette, *Strategic Approach to Addressing Technical Technician Aerospace Workforce*, White Paper describing the educational exposure/internship model developed by Boeing and the Materials Science program at Edmonds Community College (Boeing, 2011).

[ix] Edmonds Community College, "Composites Technology/Materials Science Career Pathways," (Lynnwood, WA: Edmonds Community College Relations Office, 2011).

13
Chapter

Secondary/Postsecondary Programs of Study for IT Convergence Technicians

Ann Beheler, Ph.D., Collin College

IT Convergence Technicians

Convergence technicians support the infrastructure of the converged network that simultaneously carries voice, data, video, and image traffic in a secure manner. Essentially, convergence technicians are Information Technology (IT) professionals who support the framework that provides end-to-end communications through the Internet and within businesses. Convergence technicians work with both wired and wireless technologies and also facilitate mobile device access and integration. There are various levels of jobs available to a highly educated convergence technician, and these range from entry-level positions such as help-desk and retail network support to higher-level positions such as network support specialist, sales engineer, and mobility specialist. This is a growing field, and new, innovative technologies, such as cloud computing and network virtualization, are emerging almost daily. As a result, there is little standardization in terms of job titles.

The Convergence Technology Center

The Convergence Technology Center (CTC), a National Science Foundation Regional Center of Excellence, is a growing, collaborative consortium of twenty-four colleges, distributed across the nation, that prepare students who seek education and training at both two- and four-year degree levels to prepare them for careers as IT convergence technicians. The Center was founded in 2004 as a combined business and college response to the downturn in IT employment for network technicians that occurred as a result of the "dot-com bust" and the tragedy of September 11, 2001. Working with businesses and industries in the North Texas region, the Center initially used a proven job-skills analysis process to determine the knowledge, skills, and abilities (KSAs) for potential employees that employers felt they would need in twenty-four months when the businesses projected that hiring would begin to rebound. The faculty at the colleges cross-referenced these KSAs to existing courses to identify gaps. Course designs and teaching modules were created to fill these gaps, and two types of programs were created: certificate programs that ranged from fourteen to thirty or more units, and associate-degree programs.

In the years since 2004, the Business and Industry Leadership Team (BILT) has met quarterly to advise the Center on emerging trends and update the KSAs needed for students who wish to be competitive for high-paying jobs in the IT convergence area. The colleges associated with the Center continue to have a combined enrollment of over five thousand students in this field. Each year, the Center provides "deep-dive" professional development for the faculty of these colleges and for instructors from feeder high school programs to ensure that they are well prepared to teach courses that address emerging technologies such as virtualization, cloud, and unified communication, all as specified by the BILT. This assures that students who complete these programs possess not only the core skills necessary for their work, but also the soft skills that are required to be effective in the marketplace.

National BILT

The foundation of the CTC is the active engagement and commitment of the businesses in the region to rebuild their pipeline of IT/Telecommunication workers. The BILT has organically grown to national representation. Representatives from large, medium-sized, and small businesses have consistently devoted time and energy to quarterly advisory meetings and to helping with recruitment, professional development, and teaching. These activities will continue, and employers from each of the new partners and mentored colleges have been and will continue to be invited to join the council. The CTC has expanded this council into a national forum of business leaders who have been providing direction for the CTC partner and mentored colleges to assure a more comprehensive nationwide view of job needs.

BILT quarterly meetings are largely accomplished through telecommunications and consistently include active representation from over forty regional and national businesses. A representative list of companies includes Dell, Cisco, Juniper, Citrix, EMC, AMX, HP, Alcatel-Lucent, Time Warner, Comerica Bank, Raytheon, Juniper, Telecom Electric Supply, Phillips Medical Systems, Consolidated Restaurant Operations, Inc. (manages restaurants such as Chili's), Titan Home Theatres, Texas Health Resources, Texas Instruments, Lockheed Martin, and Samsung; meetings also include representatives from the North Texas Workforce Board, InterLink, CompTIA, and several publishers. New BILT members are added each quarter.

To support curriculum design and updates, the National BILT provides a broad-based view of the job skills that IT Convergence Technicians need. Every year, affiliated businesses provide a rolling five-year forecast of their need for IT convergence technology workers. Business people see convergence education programs as essential to the creation of a prepared workforce. As a result, the CTC has an impressive list of advisors who are ready and willing to support convergence events (such as recruitment and presentations to students) around the country, for both colleges and high schools alike.

National Demand for Networking (Convergence) Technicians

The growing demand in the field of IT convergence is underscored by recent market forecasts. According to the annual forecast survey by *Computerworld* for 2012, almost 29 percent of IT executives contacted (n=353) said they would be hiring IT employees in the next year. "We're seeing strong hiring across the board, among organizations of all sizes," said Mike McBrierty, chief operations officer for the technology staffing division of Eliassen Group.[i] He reported pent-up demand for infrastructure upgrades and investments that have been shelved over the previous three years. Of the nine high-demand skills that IT executives identified, four relate to convergence: networking, data center, security, and telecommunications. According to John

Reed, executive director of Robert Half Technology, IT professionals with networking skills continue to be in high demand and have been "for a few quarters." Reed added that this demand has been fueled, in part, by virtualization and cloud computing projects.[ii] The survey results revealed that 35 percent of IT employers plan to hire workers with networking skills, and are looking for job candidates with VMware and Citrix skills. According to Gartner, Inc. (an information technology advisory company), of the top ten technology strategies, six fall into the IT convergence space and are an evolution of the convergence technology topics covered by the center: cloud computing, mobile applications, video over Internet protocol IP, storage class memory, ubiquitous computing, and fabric-based infrastructure. These topics mirror BILT feedback.

According to the U.S. Department of Labor's Bureau of Labor Statistics, "Two of the fastest growing detailed occupations are in the computer specialist occupational group. Network systems and data communications analysts are projected to be the second-fastest-growing occupation in the economy. Demand for these workers will increase as organizations continue to upgrade their information technology capacity and incorporate the newest technologies. The growing reliance on wireless networks will result in a need for more network systems and data communications analysts as well. Computer applications software engineers also are expected to grow rapidly from 2008 to 2018. Expanding Internet technologies have spurred demand for these workers, who can develop Internet, Intranet, and Web applications."[iii]

Finally, in the North Texas region alone, a recent labor-market-demand survey indicated the need for more than 1,900 convergence technicians in the next five years, and most of these positions are attainable by those with a two-year degree. (It is also important to note that many businesses hire two-year graduates with the goal of helping them attain a four-year degree through the company's educational reimbursement plan.) Also, the Department of Labor's *2010-2011 Occupational Outlook Handbook* projects that between 2008 and 2018, overall employment of computer network, systems, and database administrators will increase by 30 percent, and employment of computer support specialists will increase by 14 percent. Both of these projections are much faster than the average projected employment rates for all occupations. It is anticipated that these occupations will add 286,600 new jobs nationally over that period.

High-Wage, High-Demand Workers Need Cutting-Edge Skills

Unlike in some areas of technology, the skills needed for IT convergence technicians continually and rapidly evolve. Each year, the desired knowledge base for these technicians has expanded. Our BILT group has driven the Center to address emerging technologies such as virtualization and green IT. In recent meetings, the BILT has advised the Center that the IT convergence industry is an industry in transition, undergoing a paradigm shift more drastic than the shift that occurred during the development of the original personal computers. This feedback reflected industry trends identified by Gartner. The converged network is a connected network that is continually evolving. Meanwhile, we are seeing massive increases in the variety and types of devices people use to access this network, including smartphones, tablet computers, and iPads. Cloud computing, both public (such as that offered by Google and Amazon) and private (such as that owned by large companies) requires a different knowledge base than was needed when companies owned their own local area network. Software is delivered as a service, and even the network itself is being delivered as a service. Virtualization has expanded from desktop and server virtualization to storage, input/output, and networking virtualization—all topics the IT convergence technician must be familiar with. The CTC has processes in place to keep up with the pulse of the rapidly changing technologies of the industry, and it will use these processes to provide curriculum and laboratories with cutting-edge technologies so that students are well prepared to assume high-paying positions.

IT Certifications

The IT industry supports high-stakes certification testing, which results in various types of certifications for the individual seeking employment in this field. Employers usually highly value these certifications when they're hiring entry-level candidates because, in addition to the student's academic certificate or degree, the certification validates another level of specific knowledge. IT Convergence Technology programs across the nation assist in preparing students for such certifications as:

- CompTIA Certifications
  - ✓ A+
  - ✓ Network+
  - ✓ Server+
  - ✓ Linux+
  - ✓ Security+
  - ✓ Advanced Security+
- Microsoft Certifications
  - ✓ Windows 7
  - ✓ MCITP Server Administrations
  - ✓ MCITP Network Infrastructure
  - ✓ MCITP Active Directory
- Cisco Certifications
  - ✓ CCNA
  - ✓ CCNA+Voice
  - ✓ CCNA Security
  - ✓ CCNP
  - ✓ Cisco Unified Call Manager
- CWNA Wireless Certification
- VmWare Certifications
- Citrix Certifications
- EMC storage Certifications

Preparing IT Convergence Technnicians in STEM High Schools

High schools that provide pathways to community college programs in IT convergence technology typically offer one of two kinds of programs to their juniors and seniors. One program includes non-vendor-specific certification-related courses such as A+, Network+ ,and sometimes Server+; a more popular approach in some parts of the country provides Cisco CCNA training, four courses that directly articulate with community college IT Convergence programs and are sometimes offered as dual-credit courses in collaboration with the community colleges. These courses would take up both the technical core and the technical specialty slots in a normal high school program.

Alternative 1 - Cisco-specific version							
Soph 2		Elective	Humanities		Security +	Small Office/Home Office Capstone	IT Elective
Soph 1	Elective		Social Science		Linux	Information Storage Management	Implementing Network Infrastructure
Fresh 2	College Algebra			Physical Science	Wireless Telephony	Digital Home Technology Integration	Voice over IP
Fresh 1		College English		Physical Science	Convergence Technologies Overview	Implementing and Supporting Windows Desktop	Implementing and Supporting Windows Server
12th Grade	Algebra 2 w/Trig	English 12	Government	Physics	Health	Cisco CCNA 3-4	Programming or web technologies
11th Grade	Math Applications	English 11	American History	Chemistry	Physical Education	Cisco CCNA 1-2	Computer maintenance
10th Grade	Geometry	English 10	World History	Biology	Physical Education	Foreign Language	Principles of Engineering
9th Grade	Algebra 1	English 9	Geography	General Science	Physical Education	Foreign Language	Intro to Engr Design

Figure 13.1. Suggested Secondary-Postsecondary Career Pathways for IT Convergence Technology (possible dual courses shaded)

Alternative 2 – Non-vendor-specific version (Note that the student also needs to complete a capstone course in the second semester of the sophomore year in place of one elective.)							
Soph 2		Elective	Humanities		Security +	Information Storage Management	Implementing Network Infrastructure
Soph 1	Elective		Social Science		Linux	Digital Home Technology Integration	Voice over IP
Fresh 2	College Algebra			Physical Science	Wireless Telephony	Cisco 3 and 4	Implementing and Supporting Windows Server
Fresh 1		College English		Physical Science	Convergence Technologies Overview	Cisco 1 and 2	Implementing and Supporting Windows Desktop
12th Grade	Algebra 2 w/Trig	English 12	Government	Physics	Health	Network+	Programming
11th Grade	Math Applications	English 11	American History	Chemistry	Physical Education	Computer Maintenance (A+)	Web technologies
10th Grade	Geometry	English 10	World History	Biology	Physical Education	Foreign Language	Principles of Engineering
9th Grade	Algebra 1	English 9	Geography	General Science	Physical Education	Foreign Language	Intro to Engr Design

Figure 13.2. Suggested Secondary-Postsecondary Career Pathways for IT Convergence Technology (possible dual courses shaded)

Collin College and El Centro College offer representative community college programs in convergence and computer networking. For information about Collin College's programs, please see http://www.collin.edu/academics/programs/convergencetech.html. Information about programs at El Centro College can be downloaded at these addresses:

- http://www.elcentrocollege.edu/Program/IT/Convergence/docs/Insert%201.pdf
- http://www.elcentrocollege.edu/Program/IT/Convergence/docs/Insert%202.pdf
- http://www.elcentrocollege.edu/Program/IT/Convergence/docs/Insert%203.pdf

For more information about the Convergence Technology Center, please refer to http://www.connectedtech.org.

[i] Rick Saia, "9 Hot IT Skills for 2012," *Computerworld*, September 26, 2011, http://www.computerworld.com/s/article/358381/9_Hot_Skills_for_2012 (accessed May 30, 2012).

[ii] Ibid.

[iii] United States Department of Labor, Bureau of Labor Statistics, "Employment Projections: Fastest Growing Occupations" (December 8, 2010), http://www.bls.gov/emp/ep_table_103.htm (accessed October 19, 2011).

Commentaries on CPST: Required Changes, Issues, and Strategies

Dan Hull

Introduction

Change is never easy in education. A decision to initiate a successful plan for career pathways for STEM technicians will require many kinds of change:

- Change in curriculum
- Change in standards
- Change in school culture
- Change in perceptions about the purpose of high school
- Change in attitudes about the potential of students with average levels of academic achievement
- Change in perceptions regarding the value of technician preparation and careers

I hope that the proposed career pathway for STEM technicians will evoke positive responses from secondary and postsecondary administrators, boards, and teachers, as well as students and their parents. However, it may also incur some resistance and pose several issues that could stand in the way of needed innovation. We must overcome these barriers so that we can serve greater numbers of deserving STEM students and provide our nation the complete technical workforce that it needs and deserves.

I asked eight nationally recognized leaders in education to study the plan set forth in chapter 4, to identify the significant issues this plan evokes, and to suggest strategies and rationales for dealing with them. These commentaries speak to four themes that make up the first four sections of this chapter: Community College Issues and Opportunities, High School Issues and Opportunities, State Issues and Challenges, and Technician Students: Characteristics and Guidance. The chapter's final section, A High School Model, includes a ninth commentary from a highly innovative high school administrator who is successfully leading her school through this change process.

COMMUNITY COLLEGE ISSUES AND OPPORTUNITIES

Promoting Career Pathways for STEM Technicians within a Community College
Jim Jacobs, President, Macomb Community College

Dan Hull's call for a career pathway for STEM Technicians could not have come at a better time for American community colleges. One of the missions of the community college—workforce preparation—is under considerable scrutiny, because today's students no longer see many of the traditional occupational programs offered as attractive options. Macomb Community College, with the exception of health care programs, increasing numbers of students avoid many of the occupational program paths and instead focus on arts and science coursework in preparation for pursuing a four-year degree. This demonstrates that traditional-age students have internalized the conventional wisdom that four-year degrees translate into higher earning power. As a result, they are increasingly using the community college as an affordable stepping-stone to bachelor's degree completion versus learning a technical skill that would make them immediately marketable in the workplace. For example, in our community, where unemployment rates are above the national average and companies are looking for Computer Numerical Control (CNC) operators, enrollment in CNC programs is lagging behind industry demand.

The idea of aggregating technical skills with a strong science and mathematics orientation is an important new recipe for successful community college career programs. This approach will provide students the technical skills to start their careers, along with the science and mathematics foundation necessary to pursue advanced educational goals and career aspirations. Within the college, the development of career pathways for STEM technicians creates the opportunity to link science and mathematics classes directly to the mastery of basic occupational skills. It would allow these occupational areas to realign their curriculum as college preparation courses. This could help employers recognize the value of associate degree holders as skilled candidates for technician positions and encourage them to seek individuals with the credential. For colleges, curriculum realignment would mean new markets for occupational programs and the growth of these programs. Finally, the community served by the college would benefit from a strong pool of skilled technicians able to support local workforce needs, and this improved workforce would serve as a basis for further economic development.

The ideas contained in the other chapters of the book—if focused into a local perspective by a community college committed to serving its local economy—can become an important strategy for meeting these new demands. My discussion here focuses on how career pathways for STEM technicians (CPST) can be developed by a community college. Our work at Macomb Community College is still evolving and continues to be a part of our institutional planning process. Much of it has been done by trial and error, based on the long-term interaction we've had with local business and industry. If we had had the initial roadmap outlined by the CPST process in earlier efforts, our former efforts would have benefited enormously.

The development of CPST has significant institutional consequences that go well beyond simply instituting curriculum change in one department. The changes necessary are broad, with implications for teaching loads as well as program development. This means that support from the leadership of the institution is critical. Traditional liberal arts faculty members need to accept that *all education is career preparation*, including the college's traditional arts and sciences transfer programs. Mathematics and science programs must be integrated within technical programs. And faculty in the occupational program areas must adjust their perspective on how they teach, recognizing that their students are not only interested in a terminal degree but also a springboard to pursuing a bachelor's degree. At Macomb, with the exception of our health care programs, most "occupational programs" do not have mathematics and science requirements. This was motivated by the desire to get students quickly into the workplace. However, given today's realities in workplace skill requirements, the lack of these classes will prevent students from moving along a career pathway.

Second, it is important to realize that CPST is a general approach to developing an occupational pathway within an industry. For it to be relevant to the student, the institution, and the community, it must be focused on the labor markets that the institution serves. A key characteristic of the majority of community college students is that they normally remain within the community after completing their degrees. Thus, it is vital that the institution work with their industry partners to chart out the specific career pathways that exist within the dominant industries of their community. This requires a good deal of dense interaction between local business and industry and their community colleges on an ongoing basis. While this is not easy work and requires resources from the institutional research office, the rewards for the institution, students, and community outweigh the expense.

Macomb has undertaken this type of market investigation for some specific fields related to the auto industry. Our county is the home of the auto design industry, and over the past generation, this occupation has undergone profound change, forcing the college to adapt programs, equipment, and staff. In the 1970s, auto designers were recruited out of high school shop classes and worked on drawing boards. Next, employers began expecting completion of coursework at the college in specific areas of auto design, usually amounting to a one-year certificate. In the 1990s, the pervasive use of computers in auto design led employers to require an associate degree to ensure that designers were proficient in computer-aided design, as well as the specific software used by the different automakers. For years, an associate degree in auto design from Macomb was considered appropriate training and education for entry-level positions with the original equipment manufacturers. However, by the mid 2000s, the design function became increasingly integrated with other aspects of new model process, and large automotive companies began to seek design engineers with bachelor's degrees. The Macomb program continued to evolve, adding coursework that would position students to transfer into four-year degree programs.

Third, once a career pathway is established, it is important for the college to continue supporting the development and growth of that pathway by upgrading equipment, developing faculty, and continuously researching and monitoring the quality of the programs.

221

Given the level of activity necessary to develop and maintain relevant pathways, this means that most colleges will only have the resources to create pathways in two to three areas. At Macomb, we have focused on programs related to our region's primary industries: automotive and defense.

Fourth, for other programs, it will be necessary to partner with other institutions to share curriculum, staff, and ties with industries. At Macomb, we are attempting to develop additional career pathways by leveraging the experience and work of other institutions. We work closely with a group of thirty community colleges that are all located in auto communities, because we tend to have similar local economies and can share common approaches. It would be extremely useful for organizations such as the National Science Foundation to use the ATE Centers to form these peer-learning institutional arrangements so that colleges that serve communities with similar economic and industry profiles can share their approaches to creating and maintaining career pathways.

At Macomb Community College, we are working closely with our K–12 partners, especially middle-school- to high-school-age kids, to ignite the passion in students for the science, technology, engineering, and math (STEM) fields. Many new and exciting emerging fields encompass STEM concepts, such as mechatronics, advanced automotive technology, and green building design and construction. Our interaction with secondary students takes many forms, including early admission and special dual-enrollment programs that give high schools students a first exposure to college classes. During the summer, Macomb offers summer academies that help students learn about careers options in robotics, electric vehicles, and renewable energy. These academies introduce STEM postsecondary and career options to potential students. In conjunction with Lawrence Technological University, Macomb also hosts a Roboparade in which children of ages 8–14 create LEGO robotic floats able to follow a parade path without human intervention. A culminating year-end event is the Robotics Engineering and Technology (RET) event hosted by Macomb College, Utica Schools, and the U.S. Army Tank Automotive Research, Development and Engineering Center (TARDEC), in which middle and high school students can interact with and observe different STEM equipment. Throughout the year, Macomb's College 4

Kids (C4K) group offers a variety of fun and interactive workshops to spark younger children's interest in STEM. The key to Macomb's STEM strategy is to show children, parents, and educators that learning about STEM can be fun and exciting. When STEM learning is applied, students often do not realize that they are learning STEM concepts. Macomb offers many programs in the STEM fields that can lead to meaningful and gainful employment and articulate into baccalaureate degrees.

Finally, CPST should not be undertaken as a boutique program for a talented few, but as a means of enhancing success for all students. In that regard, steps must be taken to ensure that mathematics and science prerequisites, which are necessary for admission to some of these programs, can be mastered by students who may start in developmental education. We are only beginning to deal with these issues at Macomb. Our starting point is to learn from our other selective programs, such as nursing. Every year in this highly competitive area, approximately seven hundred students apply for 160 available slots. About 20 percent of the students accepted into the nursing program started in developmental education programs at the college. We are trying to determine the factors that made these students successful, including identifying any interventions that made a significant difference. We hope to translate these insights into our work with our ATE Center for Advanced Automotive Technologies. Because business, industry, and technology are constantly evolving and affecting the relevance of education and training, we recognize that our work at Macomb Community College is never complete. We are committed to continuing to learn from other institutions. We believe that the CPST approach also complements other national programs focused on student success, such as Achieving the Dream. We will continue to focus on helping students achieve success by helping them develop marketable occupational skills and preparing them with a strong foundation to pursue advance educational opportunities. These efforts will increase the value of the community college both for our students and for the communities we serve.

The Challenges and Leadership for Success:
Postsecondary Views
George R. Boggs, PhD, President and CEO Emeritus of the
American Association of Community Colleges (AACC)

STEM technician careers are rewarding for the individual, and encouraging students to pursue them is important for the future of our country. However, these careers are not well understood by students, parents, and high school teachers and counselors—and students are too often not prepared to succeed in STEM courses. Through concerted action, community colleges can overcome the barriers to student access to and success in these programs.

Barriers. Too many obstacles to a technological career path remain. Many students graduate from high school unprepared for college and must take remedial or developmental classes. Even if the students were generally prepared for college, they may not have taken the math and science classes in high school that would enable them to be successful in college STEM classes.

Students also have little knowledge of or perhaps inaccurate perceptions of STEM careers. Popular movies and television shows often portray scientists and technicians as socially isolated individuals who work alone in laboratories with equipment instead of with other people. Furthermore, STEM jobs are not seen as directly helping other people, in contrast to, for example, a career in allied health. Research shows that the primary reason that women are less likely than men to go into science or engineering fields is that women are more likely to want jobs that directly help people.[i] Students need better and more accurate information about STEM careers, and they need to see role models: people who look like them who have chosen STEM career paths.

Strategies. Community college faculty and administrators should get students onto campus to tour science and technology laboratories and to talk with STEM faculty and STEM students. These experiences should begin as early as elementary school and continue through middle school and high school. Colleges can motivate students by involving them in college projects or research. Colleges can arrange for students to tour STEM-related businesses to give students a more accurate perception of what it means to pursue a STEM career.

Community colleges can also sponsor science or STEM fairs and "summer camps" to encourage students to work on projects that engage their interests. High school students can enroll in community college classes through dual or concurrent enrollment programs. Some community colleges have early-college high schools located on their campuses, which provide even greater opportunity to engage students in STEM projects.

Articulation with local high schools is important. Community colleges have started administering college placement examinations to students while they are in high school and still have time to take the courses they need to adequately prepare for college. Colleges should also be sure that high school counselors and teachers are invited to campus to tour facilities and to interact with their counterparts at the college.

Community colleges are becoming more engaged in teacher preparation and professional development. Colleges that have instituted professional development programs for teachers have opportunities to engage them in STEM projects and research that build knowledge and understanding that can translate into greater enthusiasm in their elementary, middle school, and high school students.

Parent orientations are also becoming more common on community college campuses, and they provide an opportunity to educate parents about the career pathways, including those in STEM fields, that their children can pursue.

Bridge programs can make a big difference in students' success, especially students from low-income families and minority populations. These programs assist students in their transition from high school into college. Information about STEM programs and STEM careers can easily be integrated into these bridge programs.

Recommendations for trustees and presidents. Encouraged by initiatives such as *Achieving the Dream: Community Colleges Count*, trustees and presidents are "creating cultures of evidence" on community college campuses (see http://www.achievingthedream.org/). Planning and decisions are informed by data, and reports on outcomes of programs are now more prevalent. Certainly, if the top policy makers and the chief executive of the college believe in the importance of a program or an initiative, and if they ask for regular reports, it is a clear signal of the importance of that program or initiative.

Examples of data that can be collected and reported include: what percentage of incoming students are unprepared for college or unprepared for college STEM programs; what percentage of students in developmental classes are successful; what percentage of students in STEM technician programs complete their programs; whether good jobs are available locally for STEM technician graduates; and whether the college's programs are preparing enough graduates to meet the needs of local businesses. Information that is gathered should be disaggregated by gender, ethnicity, and economic status of the students and by which high school they graduated from. Trend data can inform the college community, employers, and local high school leaders about the extent to which strategies for improvement and strategies to close achievement gaps are effective.

College presidents can work to improve relationships with local schools by scheduling regular meetings with school superintendents and principals to discuss issues and strategies for improvement. Trustees can meet periodically with their counterparts on local school boards to discuss policies that remove barriers for students. As noted in chapter 4, high schools can implement an alternative curriculum for students who are not currently being well served. Dual-credit policies can give students a successful launch into college.

Recommendations for vice presidents and deans. While the trustees and the presidents should focus on measuring progress toward outcome goals, it is up to the administrators in instruction and student services to work with faculty and staff to develop strategies designed to improve the outcomes. Promising strategies to close achievement gaps and increase completion rates include the use of learning communities, study groups, peer tutors, study skills classes, freshman experience classes, bridge programs, parent orientation programs, and counseling, as well as the elimination of late registration for classes.

Recommendations for faculty and counselors. Counselors and STEM faculty members are the key to closing the achievement gap, improving college completion rates, ensuring that students have the information and support they need to find the programs that they are suited for, and helping students succeed. Regular meetings with industry advisory committees and other employer representatives can help ensure that the curriculum is up-to-date. Inviting teachers and counselors to the college for tours and informational meetings will give them an opportunity to learn more about college programs and the quality of instruction. Getting students onto campuses and into local businesses can inspire them to continue their education and give them the desire to enter a STEM field. Establishing local science fairs or summer science programs can give students the confidence they need to pursue a STEM-related major when they enter college.

Conclusion. Previous chapters make clear the challenges we face to maintain a strong and vibrant society, how important STEM and STEM technician careers are to our future well being, and how rewarding these careers can be. However, today's educational structures are not serving us—or our students—well. It is time for us to implement strategies that will overcome the barriers and data systems to monitor our progress.

---

[i] Jacquelynne Eccles, "Who Am I and What Am I Going to Do With My Life? Personal and Collective Identities as Motivators of Action," *Educational Psychologist* 44, no. 2 (2009): 85.

HIGH SCHOOL ISSUES AND OPPORTUNITIES

Form *Should* Follow Function

J D Hoye, President, National Academy Foundation

The largest challenge to bringing industry-valued and -validated curriculum to schools is our mind-set about the purpose of education. Industry, political, and educational leaders readily invoke the catchphrase "college and career readiness" and follow it with laments about our inability to prepare enough students to meet that standard. The need for curriculum, instruction, and high school design to be aligned and organized to prepare students for emerging careers is obvious. So why are such alignment and coordination not common practice?

First you need to understand what drives high school design and curriculum choices. Sadly, in most cases, it's not assuring that graduates are career ready or equipped with skills for the workplace. Instead, they are predominately driven by traditional academic structures that inhibit connections to workplace preparation and connections between the disciplines themselves.

Making the shift to assuring that academic preparation and career preparation do not compete but enhance one another is the magic. As with designing a product to solve a problem, being clear on the function expected before devising the form that it will take is essential to success.

In order to bring high schools relevant curriculum that bridges to postsecondary institutions and is valued by industry, we must:

1. **Stop the fight about which is more important: academics or technical skills.** Integration is the answer. Too often, courses are taught in such isolation that no connection is made by the learner and leveraged proficiencies are harder to come by. The systems challenge is in part tied to how the school day is designed, with its overly prescribed use of time and rigidly constructed periods that separate subjects from one another. We can address this design challenge by emphasizing common planning time for teachers and by grouping students into

cohorts. Integration cannot be adequately addressed after the schedule is planned. Shared planning time gives focus to how both content and instructional approaches are organized so that students see the relationships between disciplines, workplace skills, and future career choices. Not only can this be accomplished without compromising academic quality, in fact, it acts as a curriculum unifier. There are many intermediaries that can help with project-based design needs, such as the Buck Institute for Education and ConnectEd: The California Center for College and Career.

2. **High-quality teachers are essential.** Teachers must have an understanding of STEM content in a career context as well as in an academic one. This can be accomplished in pre-service and in-service trainings. In both cases, a robust connection between teachers and the workplace can transform instructional practice to more fully integrate essential knowledge, skills, and abilities into the course content. One powerful, underused approach is the teacher-industry externship. While many teachers come from industry, there are many more who come directly to the classroom. Externships offer teachers insight into how content knowledge is applied in the workplace. They create the relationships between schools and employers that can result in work-based learning experiences for students.

3. **Course content must be continuously refreshed.** Unlike many academic courses, in which revision means a textbook gets a new cover or is migrated online, technology-based courses must be continuously maintained and refreshed since the field evolves so rapidly. Intermediaries can assist local schools in this effort, reducing the burden and duplication of individual schools' efforts. Project Lead the Way, STEM Academy, Paxton/Patterson, Cyper Watch and the National Academy Foundation all provide curricula in which technical courses are kept up-to-date, are framed in the context of the industry, and use proven strategies to integrate subject matter in a project-based instructional format. All these curriculum designers, and others like them, validate their content with industry and postsecondary institutions.

4. **Let practice drive policies.** If you believe that you get what you measure, then make certain that college and career readiness is more than a mantra. It must have indicators that student performance is measured against and that are aligned with the content provided. Guidance will come from state and federal agencies, but the specifics must be clearly defined locally in order for instructional practice to result in the desired student proficiencies. Likewise, more sophisticated and balanced assessments are needed, and they must move beyond just test taking. Now, more than ever, it is time to measure what students know and can do in projects that replicate the world they will enter as they complete their formal education. We should pay an equal level of attention to performance and the ability to apply knowledge as we have paid in recent years to testing the retention of facts and knowledge acquired. In addition, policies that facilitate structures such as cohorting, shared planning time, workplace learning, and early college need to be in place. A key policy objective is to move high schools more toward a competency design rather than being designed around seat time. Such a change would enable creativity in the use of time.

   Other policy changes that would promote using the workplace as a learning place are mechanisms for students to earn high school credits for workplace learning and financial incentives for employers to provide workplace-learning experiences for high school students (with safeguards to prevent using interns to displace full time workers).

5. **Relationships matter.** Teachers and students need authentic connections to industry professionals to reinforce the skills that employers require and to provide opportunities for students to demonstrate those skills in both the workplace and school. Building the capacity at the local level to offer workplace-learning experiences is a challenge. It is best accomplished in partnership with local organizations and by a formal advisory structure made up of community leaders who can make it a priority. To formally connect this life-changing experience to perceived value by institutions of higher education and the community at large, it needs to be validated by an assessment

and be credit-bearing. Without these, it will continue to be seen as an extra and not an essential. It is particularly important for schools in low-income communities to build these relationships because students in poverty are more likely to lack these connections through family and friends.

For the success of our country, our citizens, and our students, we must acknowledge that the high school years are a critical time for introducing young people to emerging careers that require specific technical and broad academic skills. Students can then enhance their skills in postsecondary educational settings and industry. Our historical strength as a nation is our belief that all students have a fundamental right to a good education. To avoid limiting students' options, we have resisted the approach taken by some other countries, in which high school students are forced to choose between college and careers. The knowledge economy offers us an opportunity, as well as a requirement, to align and integrate what is required in college and what is necessary in the workplace, assuring the maximum set of choices for our children, but demanding a much more flexible and connected educational system.

———————————

STEM High Schools and the Technician Level: Issues

Kathy D'Antoni, EdD, Assistant State Superintendent
West Virginia School System for Career Technical, Adult, and
Institutional Education

The nonexistence of the "middle" student in STEM related pathways is a reality. Students participating in STEM pathways and activities are the top 10–20 percent of academic students. The disparity or the lack of "middle–student" participation can be attributed to a number of diverse factors, and each one is a significant barrier.

Issue One: Perception

Through communication and publication media, perceptions of STEM have evolved. STEM is now perceived as an academically focused curriculum in science, technology, engineering, and mathematics—an advanced-placement curriculum geared to higher-achieving students. Rarely, if ever, do you see STEM pathways that clearly depict various entry and exit points leading to technician-level positions. In education circles, the acronym STEM is now considered the "elite curriculum," intended only for the best and brightest. This perception contributes to limited access and accessibility and encourages selectivity of potential students. This is manifested in the counseling arena during the scheduling of student classes and pathways. As is seen across the states in Career Technical Education (CTE), counselors or advisors in the educational arena do not fully understand or academically appreciate the knowledge and skill sets required for highly paid, highly skilled employment as a technician. This phenomenon has existed since the 1950s and plays a major role in the United States' inability to hold its lead in educational attainment.

Solutions:

- Counselors are paired with a business or industry team and participate in internships or seminars on a regular basis, away from the school and in the workplace.
- Develop a technician pathway that is adopted or sponsored by a business or industry. The business or industry works hand in hand with the school system to develop the curriculum and provide mentors and work-based learning experiences.
- Find a new acronym for the technician pathway— a "sexy," catchy label that will pique the interest of young students.

Issue Two: Curriculum Design

Students learn in different ways, and research shows that the best and most efficient learning takes place when students are engaged. The reason CTE has been so successful in teaching students is because of the hands-on learning that takes place in the classroom. STEM activities that are hands-on must be introduced in the elementary curriculum. Middle school is too late to garner student engagement and interest in the STEM-related fields; by that time, students have self-selected out of "that type" of curriculum based upon the kinds of misperceptions discussed above.

Solutions:

- The elementary curriculum needs to embody the foundational skills found in the STEM arena.
- Build the elementary and middle school curriculum with the end in mind. Clearly depict the foundational skill sets needed for success in STEM careers, and then backfill the curriculum with project-based learning and hands-on activities at each level to create the pathway.
- The body of knowledge within the curriculum framework in our schools needs to be developed in learning modules or skill sets. Chunking the curriculum in this manner enables easy identification of remedial needs, allows all students the opportunity to achieve mastery, and succinctly identifies for students what they need to know and be able

to do, which places equal responsibility for achievement on the student.

Issue Three: Career Development

"All my life I wanted to be somebody, but I see now I should have been more specific."[i] This statement is holding true for a large percentage of today's youth. Today, the fastest-growing major on our college campuses is the "undecided"; students are taking six to eight years to graduate from college with a four-year degree, if they graduate at all; and far too many of our young adults are coming to the realization that they have neglected to obtain the necessary academic and technical skills to be successful. High school students view the high school curriculum as a laundry list of courses they must take to graduate, and many fail to see the connection with life after high school. As our country moves further into the information age, it is important for the youth of America to become more specific in their career preparation. Career pathways can provide that specificity.

History shows that merely raising standards and setting benchmarks to reflect increased rigor and expectations for students will not yield the intended results. This reality has been observed consistently over time, and yet we continue to repackage and reintroduce this flawed concept in public education.

The solution lies in restructuring the delivery, defining purpose in the curriculum, and delivering instruction that actively engages students, empowers them to act, and empowers them to assume greater responsibility for their own learning. In other words, relevance should precede rigor and increased expectations.

A major deficiency in the current middle school, high school, and college delivery systems relates to the lack of a quality, equitable, career guidance program that prepares students to make informed decisions and set realistic career and education goals. The lack of such a program results in students making decisions concerning what to take in school, how hard they should work, or their post high school plans without accurate information. The result of this deficiency is a large number of students in an unchallenging "default" curriculum, disengaged, unmotivated, and frustrated. This, in turn, translates into high course-failure rates, low state and national test scores, high dropout rates, low college-retention rates and high youth unemployment.

Whatever the college or career path chosen, it is certain that the choices students make will become increasingly important. Yet, the public schools have often disinvested in the guidance function at the same time that the available counselors are often diverted to administrative and other non-guidance functions.

Solutions:

- A comprehensive career development system needs to be incorporated from kindergarten through twelfth grade. The system needs to encourage, support, and provide investigative opportunities for all students that will provide them the necessary career-information skill sets that will allow for quality decision making about their future career.
- The culture of the school must be focused on providing a career-awareness arena for all students. Career development cannot be the responsibility of just the school counselor; it must be the responsibility of the school as a whole throughout K–12 education.
- Career pathways must be clearly identified and articulated.

Issue Four: Quality and Equity

The high school diploma can no longer be viewed as the culminating education credential for future successes. The shift in the nation's economy from an industrial age to an information age requires students who seek quality employment and a successful future to continue their education beyond high school. National statistics report that at least 80 percent of students must complete postsecondary coursework.

A large number of students enter school with major disadvantages, and the achievement and attainment gaps continue to grow as these students progress through the system. Today, everyone is talking "assessment and standards," while equity discussions are few and far between. It is important to keep both quality and equality in mind. We need a set of *equity linkages* that will help reduce the effect of economic and social differences, and a set of *content-based linkages* that will smooth students' passage through the system and allow them to attain the highest levels they are capable of achieving.[ii]

Another issue is not all students have the opportunity to access college courses offered at the high school level. Currently, most school systems and colleges limit student participation, and the requirements are frequently set by subjective and resource determinations. Standards do need to be set; however, the question that arises is standards placement. Should standards be placed at the beginning of a process, thereby limiting access, or should they be applied at the end of a process as benchmarks for appropriate mastery and assessment levels? In an equitable education system, standards are primarily placed at the end of a process.

There are advantages to not limiting student enrollment in college-credit courses to honor students only:

- Average to below-average students often have easier access to support services, such as tutoring and one-on-one supervision, at the secondary level than at the postsecondary level.
- More high school students are exposed to higher educational standards and more sophisticated curriculum.
- First-generation college students gain "college knowledge," such as information about admissions requirements and financial aid.
- It sends a strong signal to all students that they have the opportunity and ability to continue their education at a postsecondary institution.

Our nation is not and will never be one-size-fits-all. As the old saying goes, there are many ways to skin a cat. So we need to institute diverse learning arenas that will allow the next generation of good students the ability to do great things.

Solutions:

- Postsecondary educational options need to be open and encouraged for all students.
- The senior year of high school should be a blend of secondary and postsecondary courses.
- Collaborative high school and postsecondary career counseling for students.

[i] Jane Wager, *The Search for Signs of Intelligent Life in the Universe* (New York: HarperCollins, 1986).

[ii] Harold Hodgkinson, All One System: A Second Look (Washington, DC: Institute for Educational Leaderships, 1999), 8.

STATE ISSUES AND CHALLENGES

Assuring the Appropriate Academics for CPST: Using Common Core Standards
Dr. June Atkinson
Superintendent of Public Instruction, North Carolina

Adopted by forty-five states, the common core standards in mathematics hold much promise for developing a strong foundation for STEM careers, regardless of the level of a student's career aspirations. These standards are based on eight guiding principles that require K–12 students to use appropriate tools to solve problems, model with mathematics, and make sense of problems. The standards reflect the need to develop a deep understanding of mathematics, as required by many high-achieving countries. They are based on the notion that fewer topics taught in a deep, rich way will better prepare students for success than will briefer exposure to a greater number of topics. The use of these standards has many advantages for states, curriculum developers, and articulation work between K–12 education and community colleges and universities.

If asked the question, "Do students in North Carolina have the same need to learn how to use math as students in California?" most people would answer "yes." If asked, "Should the standards for learning how to add, multiply, subtract, divide, use fractions and percentages, and use algebraic formulas to solve real world problems be different based on where you live?" most people would say no. Therefore, chief state school officers and the National Governor's Association worked together to develop standards that could be voluntarily adopted by any state. Some of the foremost experts in mathematics were a part of the development team. All states were asked to give feedback about the standards, which were benchmarked with other countries whose students demonstrate high levels of achievement in mathematics. While there may be disagreement about grade-level placement, wording, or a particular standard, it is a giant step forward to have these standards in place.

States have already begun to share resources for teachers to use in the classroom, as well as resources for professional development both online and in face-to-face workshops. With states sharing resources, there is the opportunity for much richer, personalized assistance for teachers across the nation. Quality assessments to be used for formative, interim, and summative purposes can be developed, as has been initiated by the SMARTER Balanced Assessment Consortium (SBAC) and Partnership for Assessment of Readiness for College and Career (PARCC) consortia, which more than forty states have joined.

Educational developers now have the potential to provide better and richer material. Instead of companies developing materials to meet the different standards of fifty states, they can now concentrate on the common core standards.

States have constantly wrestled with the question of what it means to be career and college ready. By using the common core standards as a backdrop, community colleges, universities, and state governments have a place to start that discussion. Getting agreement between public schools and postsecondary institutions in each state about the common core will help further the goal of unnecessary duplication of course work for students. It will also allow the same types of assessments to be used to determine both exit from high school and entry to postsecondary schools.

In North Carolina, the General Assembly recently passed Career and College Promise, a program that offers clear, focused, affordable career pathways to high-school-age students. Since then, the Department of Public Instruction and the Community College System have identified STEM Pathways and a career-technical pathway that give high school students a jump start by enabling them to take community college courses while in high school if they meet certain college-readiness criteria. Once students have met those criteria, they can take the required courses in math and other subjects that will lead to an associate degree or other credential. This initiative will ensure that more students take the community college courses, including math, that they need to get an associate degree, instead of taking courses that don't lead to a credential.

While North Carolina has a long, successful history of statewide articulation agreements and work between public schools and community colleges, North Carolina leaders believe that the Career and College Promise program will save taxpayers and students dollars while preparing more students for STEM careers.

The common core standards work was instrumental in facilitating the Career and College Promise initiative.

More information about the common core standards is available at http://www.corestandards.org/assets/CCSSI_Math%20Standards.pdf

---

Why the Time for Creating Career Pathways for STEM Technicians is *Now*

Willard R. Daggett, EdD
The International Center for Leadership in Education

When Dan Hull asked me to offer some thoughts on this new book, I was, of course, honored to be asked. I have known and worked with Dan for many years and know him to be a tireless advocate for our students' future needs—and therefore our national interests. He and I hold passionate beliefs about education in the context of society—beliefs that are articulately expressed in *Career Pathways for STEM Technicians (CPST)*. Dan's vision closely parallels what I share with audiences across the country and with the partner schools that the International Center for Leadership in Education has supported since 1991.

I have worked closely with senior state leadership in states such as Georgia, which enthusiastically embraced Harvard's "Pathways to Prosperity" report and supported career pathways systems in its school districts; and I have observed the promise of STEM programs in many of the states, districts, schools that my colleagues and I work with. However, it would be wrong for me to claim deep expertise in STEM, photonics, biotech, or the specific technician positions that American industry so desperately needs to fill. So I'll leave subject-matter expertise in STEM to Dan and others and, instead, share some related and, I hope, relevant and related thoughts.

I started my career as a high school teacher and college professor, and then worked in the 1980s for the New York State Department of Education (NYSDE) as Director of its "Futuring Project." The premise of that initiative was that education exists in the larger context of the society it serves, and that when that context changes, education must change too if it is to remain relevant. We recommended that leaders in K–12 education consider students' future lives, beyond high school, as lifelong learners, citizens, consumers, family members, and productive members of a skilled workforce. I would like to think that the recommendations that we presented to NYSDE anticipated in some small fashion the intent and design of the rigorous and relevant career pathways programs for the "middle two quartiles" described in *CPST*.

So I hope that the following four points will help frame some of the key recommendations for CPST. They are themes that I regularly have the opportunity to discuss with educators, business leaders and other education stakeholders across America.

1. **America needs new approaches to K–12 learning** that are better aligned with the competitive, interconnected, and digital world in which our students will learn, live, and work.

   - Our nation's future depends on our ability to develop a wealth of versatile, productive, and technically skilled talent to meet the increasingly sophisticated demands of an increasingly complex workplace in the face of growing global competition.

   - America needs increasingly more well-trained four- and especially two-year graduates from STEM-related college courses and training programs to fill the abundance of technical and technician jobs that are going unfilled now and that threaten to remain unfillable in the future.

   - The infrastructures, systems, credentials, curriculum, and instruction in our traditional K–12 system are generally not aligned with the requirements of the twenty-first-century workplace or the needs or personal lives of our students.

2. **Schools should leverage The Common Core State Standards** (CCSS) developed by the National Governors Association Center (NGA) and the Council of Chief State School Officers (CCSSO) and first released in 2010, which appear to recognize those disconnects between future needs and current curriculums and practices.

   - The CCSS build upon "the highest, most effective models from states across the country and countries around the world," but they also vary from most current states' standards in significant ways. Many have described them as "fewer, clearer, and higher," which I believe is a huge step in the right direction.

   - The Common Core State Standards also:

- o align—as their subtitle *"for College and Career Readiness"* intentionally indicates—with both college and workplace preparation;
- o are clear, understandable and consistent;
- o include rigorous content and *application* of knowledge through high-order thinking skills;
- o build upon strengths of current state standards and those of other high-performing countries to better prepare our students to succeed in a competitive global economy;
- o demand more in response to the changing demands of literacy, applied mathematics, technology, adult roles, the workplace, and postsecondary learning in the twenty-first century. They put more emphasis on learning how to apply knowledge instead of merely acquiring knowledge.

   (Source: http://www.corestandards.org/)

3. **Supportive instructional and assessment practices** need to be learned, enhanced, practiced, and adapted by teachers to create alignment with new standards and learning expectations, in particular the CCSS.

- Relevance makes rigor possible. Learning that is applied, engaging, and purposeful helps students recognize, aspire to, and attain higher levels of achievement and success.
- Relevance is nurtured by relationships. Career academies, small learning communities, and career-pathways programs cultivate supportive relationships and mutual interests between and among teachers and students.
- For most students, academically focused, traditional, abstractly designed "university-prep" English language arts, mathematics, and science courses are not the answer. What the vast majority of students need is more *applied* coursework, including:
   - o applied literacy—for example, reading and writing informational (per the CCSS, "dense") text, giving directions, and communicating in the workplace;

- applied math—for example, problem solving, estimating, mathematical modeling, conducting data analysis, using applied algebra, and predicting; and
- applied science—for example, workplace physics, biotech, healthcare, mechanics, etc.

Applied coursework is especially helpful when it is enabled with technology tools, integrated across disciplines, and made an integral component of career-technical, STEM, career-pathway, and other content-area coursework.

- Traditional teaching needs to give way to "facilitating learning." Teachers need to recognize new instructional practices, roles, and responsibilities, not as mere providers of knowledge in the Google age, but as facilitators, motivators, and enablers.

- Lectures, assigning textbook exercises, and "sit-n-git" must give way to instructional strategies that promote active learning: demonstration, modeling, inquiry, cooperation, brainstorming, role playing, real-world project-based learning, interning, job shadowing, and exhibiting.

- Assessment must be adjusted, too, to move beyond practices that only require students to demonstrate command of already-known information, process skills without context, and "get *the* answer." Assessment needs to include demonstrations of higher-level rigor and relevance—in other words, the ability to *apply* rigorous knowledge. Such demonstrations might include portfolios, projects, presentations, modeling, industry certifications, and participation. Assessment also needs to become increasingly formative, not merely summative.

- Teachers and instructional leaders need to be encouraged, trained or retrained, and supported to master these new approaches and attitudes. They need to be given the time and support to reflect and collaborate—with peers, with their postsecondary counterparts, and with business and industry stakeholders.

4. **We must view the bigger picture** both of learning and of student needs. *Career Pathways for STEM Technicians* takes a

comprehensive approach to the why, what, and how of improving the education and "adulthood-prep" experience for the vast majority of our students. It looks at education within a society-wide context and the context of preparing the whole child. For example, *CPST* recommends a focus on "soft skills"; dual-credit courses; technology skills; personal, recreational and career interests; "a broad curriculum base with options"; career-related experiences that complement coursework; and so on. My colleagues at the International Center and I also have a similar holistic vision of what every student's learning experience needs to include.

Our approach was developed during a multi-year study that the International Center co-conducted with the Council of Chief State School Officers (CCSSO) that was sponsored by the Bill and Melinda Gates Foundation. Our purpose was to identify exemplary practices used by rapidly improving high schools—traditional high schools, career academies, STEM schools, and other structures. Part of what we discovered is now captured in a framework/school analysis and support tool called *The Learning Criteria to Support 21st Century Learners.*

The Learning Criteria measures school effectiveness in four dimensions:

- **Foundation Learning. Achievement** in the core subjects of English language arts, math, science, and others identified by the school.

- **Stretch Learning. Demonstration** of rigorous and relevant learning beyond minimum requirements, such as participation and achievement in AP courses, specialized courses, career pathways and STEM programs, and so forth.

- **Learner Engagement.** The extent to which students are motivated; involved; have a sense of belonging and participation at school; and have supportive relationships with teachers, other adults, peers, and family.

- **Personal Skill Development.** Measures of personal, social, service, leadership, and workplace readiness skills, and demonstrations of positive behaviors and attitudes.

Schools use *The Learning Criteria* to self-assess, to plan, and to measure their progress in creating a balanced approach to student support and growth. Its intent clearly reflects the breadth and depth of the STEM career pathway programs described in *CPST*.

In conclusion, let me say that I share and applaud Dan Hull's leadership and enthusiasm for STEM Technician Career Pathways initiatives. Such necessary, doable, and seemingly practical innovations are what will make K–12 education more responsive to society's needs. They will also help us do what is right for the students we serve.

———————

TECHNICIAN STUDENTS: CHARACTERISTICS AND GUIDANCE

Technician Education is Different, and So Are Most of the Students
Darrell M. Hull, PhD
Department of Educational Psychology, University of North Texas

Education, career development and employment success for technically oriented students could be enhanced if educators, administrators and counselors considered the following suggestions, together with some of the implications of research on practice:

1. **The baccalaureate degree is a realistic predictor of career success for less than 30 percent of high school students.**
    The four-year baccalaureate degree is perceived as something everybody should want, but it may be a misleading path for the majority of American high school graduates. The baccalaureate degree has become esteemed by both parents and high school graduates because it is seen as the ticket to a high-wage job, and the majority of employers have endorsed the four-year degree because for them, it is a no-cost means of screening for high school graduates that are persistent and have some academic ability in subjects requiring math or verbal aptitude. Many employers refuse to grant interviews to candidates unless they possess a baccalaureate degree, further reinforcing the idea that a baccalaureate degree is necessary for success. Moreover, as more people attempt a four-year degree, it makes sense for employers to use the baccalaureate degree as the limiting requirement for an interview (despite the fact that employers probably know that there are exceptions). Consequently, the baccalaureate degree has become something almost every student is supposed to want. But at the same time, the accomplishment provides little tangible benefit to many baccalaureate degree holders. Most people who do not opt to attain a four-year degree are socially and occupationally punished.[i]

Because employers have relied on the baccalaureate degree as a minimum requirement, they have screened out thousands of potentially valuable workers. Many students who complete the four-year degree to get their ticket to corporate America are often relegated to jobs in which they find no intrinsic satisfaction because the job requirements don't fit their proclivities. Almost 50 percent of high school graduates go on to pursue a four-year education, and perhaps 30 percent of those that successfully complete the degree lag behind the top performers in earnings, are confined to a cubicle for the majority of their career, and end up hopping from position to position accompanied by minimal incremental increases in wages throughout their career. The result is a life of mediocre productivity because of decisions made at the age of seventeen.

The satisfaction of being good at what one does for a living (and knowing it), compared to the melancholy of being mediocre at what one does for a living (and knowing it). This is another truth about living a human life that a seventeen-year-old might not yet understand on his own, but that a guidance counselor can bring to his attention. Guidance counselors and parents who automatically encourage high school graduates to enroll in a four-year college, regardless of their skills and interests, are being thoughtless about the best interests of young people in their charge. Even for students that have the academic ability to succeed in college, going directly to college may be a bad way for them to discover who they are and how they should make a living.[ii]

Over the last half century, America has redefined what it means to be "academically talented," such that aptitude in two very limited cognitive domains, verbal and math (the primary domains required for success in a traditional baccalaureate degree program), are valued beyond any other cognitive abilities, including visual perception, auditory perception, memory, and cognitive speededness. The College Board, producers of the SAT entrance examination, conducted a study of forty-one colleges to determine the relationship between SAT scores and college readiness, defining readiness as a 65 percent probability of getting a 2.7 grade point average during the freshman year. The benchmark scores were 590 for the SAT-Verbal and 610 for the SAT-Math. How many American seventeen-year-olds can meet these benchmarks? The answer is between 9 and 12 percent, depending upon how one estimates, but realistically, approximately 10 percent are likely to earn a solid B average in a four-year college. The reason for these low estimates is that the test really fits well with the curriculum offered in a true four-year baccalaureate-degree program that emphasizes a required core of coursework that includes literature, composition, foreign language, American government or history, economics, mathematics, and natural science. While these are laudable subject areas for the elite in the areas measured by the SAT, they are not practical (as core baccalaureate-degree subject areas) for most employment.

The fact of the matter is that a four-year degree isn't for everyone, nor should it be. Consequently, far too many people attending a four-year college are enrolled for the wrong education, and perhaps for the wrong reasons. Worse yet, the overemphasis on math and verbal skill development using benchmark subjects such as calculus or literature is preventing America from developing talent in other areas. America has always held a college education in high esteem, but during the first half of the twentieth century, there were lots of rewarding jobs that a person could pursue without a four-year college education. Today, the landscape is much different. If parents mention to their friends that their son or daughter has decided not to go to a four-year college, the most frequent response is one of horror, as if the four-year college education is part of some magical formula for success. But, as I have pointed out, it may not be relevant, it often results in occupational dismay or depression, and it certainly does not support the development of a wide variety of abilities in a workforce that would strengthen the American economy.

2. *Individual differences* **in interest, social behavior, and learning abilities should guide students in the selection of postsecondary education and career preparation.**

Perhaps the observation we most notice about each other is that we differ from one another, sometimes in subtle ways, but also in ways that make us very distinct. This simple observation is known to psychologists as *individual differences*, and is the subject of considerable research. Individual differences are important because people perceive things differently and, consequently, behave differently. Some of these differences impinge directly on how people go about work, how they select their careers, and what they find interesting in school. People with different attitudes respond differently to direction, and different personalities interact differently with bosses, coworkers, subordinates, and customers. Some workers or students learn new tasks more quickly or effectively than others, and people usually perform well in assigned tasks that suit them.

Much of the work of psychologists in individual differences research has been to categorize the infinite number of ways we differ into a limited set of taxonomic categories often referred to as *unobservable* or *latent constructs*. These categories generally fall into two broad domains: personality and cognitive ability. There is also a conative (motivational) domain that I do not address here. Much of who we are as individuals, what allows us to distinguish ourselves from one another, is reflected in our personality and cognitive ability *traits*. Traits are relatively enduring characteristics of who we are that are largely, but not entirely, determined by our genetic makeup. One reason that traits are used to distinguish individuals in individual differences research is that traits tend to remain stable over time. Educational psychologists have extended knowledge of individual differences to groups of individuals (say, for example, males vs. females or different ethnic groups) to understand how such groups retain distinct differences after averaging together all those within a group. It is important to point out that simply because groups manifest differences between one another on some trait does not mean that an individual within that group will necessarily possess the same distinct difference or characteristic.

A related field of psychological study has to do with how these trait-based differences manifest over time and how the environments we choose for ourselves tend to reinforce or cancel out these differences. Sandra Scarr conceptualized human development as individuals finding out who they are and, concurrently, becoming more uniquely themselves.[iii] Rather than being a passive agent who simply reacts to an environment, each person is endowed with a unique personality and set of abilities and penchants that interact with the environment to produce various life outcomes—both more and less successful. From Scarr's framework, people who self-select into any career path (technical or otherwise) do so for various reasons that primarily have to do with individual cognitive ability and personality characteristics. Based on this theory, groups of people that choose a technical career path versus a nontechnical career path might vary, or demonstrate individual differences, endemic to their basic cognitive and personality traits.

Given the tendency of individuals to select life paths based, at least in part, on their skills and interests, it is an empirical question whether there are similarities in cognitive ability and personality characteristics of students that choose technological education programs at community colleges and students in other workforce programs at community colleges that are less technologically oriented.

3. **Technological education is a unique form of learning and instruction.**

Learning styles is a term that refers to the extent to which different forms of instruction are more effective for different individuals. One purpose of this commentary is to point out that instruction in technological education programs at two-year colleges is, in many ways, vastly different from that which is provided in other forms of higher education or in most traditional high school programs, where the emphasis is on liberal arts preparation. Moreover, early results from my own research reveal that students in technological education programs exhibit real differences from other students in their cognitive abilities and personalities.[iv] I suggest that students in these programs have self-selected to participate in technological education (whether they intended to or not) based on their abilities and personality. I suggest that this is a result of the unique learning environments offered in technological education: technological education offers a considerably greater focus on laboratory work to emphasize principles and does much more with schematics that organize content spatially.

Before continuing, I should clarify my position on learning styles and person-environment fit. It is one thing for someone to knowingly or unknowingly identify a learning environment that suits his or her learning style. It is quite another to suggest that learning environments should adapt to individual students—a prospect that I feel is unproductive, especially for technological education. For two decades, assessments of learning styles have typically asked respondents to report the ways they prefer to receive information or the kinds of activities they find most engaging. A great deal of activity is associated with the use of measurement and classification devices that categorize learners based on their *style of learning*. Much of this activity is supported by the popular notion that all people have the potential to learn effectively and easily if only instruction were tailored to their individual learning style. Only recently have advances been made in producing a self-report questionnaire for learners that possesses documented reliability and validity evidence for such categorization (verbalizers, spatial visualizers, and object visualizers).[v] The importance of learning styles for educational practice is less clear,[vi] as *no methodologically strong studies have been conducted to provide evidence that matching different forms of instruction to learning style produces enhanced educational outcomes.*

The reason for exploring the possibility of these differences is that if group differences exist, then future studies could explore ability-based differences in learning. Almost no research currently exists about how or whether differential abilities translate into different learning strategies (or styles of learning[vii]). Also, if verified, future studies would be able to point out whether such ability differences resulted in career satisfaction or earnings differentials that would support the assessment of individuals prior to enrollment in programs to determine if they have a greater likelihood of success. Such information would be helpful for not only counselors, but also faculty in technological programs; they might more successfully teach their students if they better understand different students' limitations. Finally, if such differences exist, theories about career development have greater utility. Such differences might be useful in designing curricula and instruction, recruiting, selecting students for programs, and matching students to careers that would offer the opportunity for a lifetime of satisfaction, particularly in contrast to other less-technologically-oriented students.

There may also be students in technological education programs who do not possess traits that technicians typically exhibit, and it may be helpful to identify these students and ways to support them. A minority of students in technological programs at community and technical colleges likely do not possess strong spatial talent. How can we help educators serve these students? While researchers frequently examine verbal and fluid (i.e., problem-solving) abilities when examining students and professionals in STEM fields, they often overlook spatial ability.[viii] To date, the few studies that have examined the educational and occupational significance of spatial ability for STEM domains have done so only with participants with exceptionally high ability (e.g., top 3 percent[ix]). At least in advanced STEM fields, spatial ability has proven to be a highly salient psychological characteristic that is predictive of success in STEM. In addition to traditional cognitive ability measures, personality factors also appear to be essential to understanding how education and work are linked with interests and occupational selection[x] and more specifically for STEM occupations.[xi]

4. **Technical students' spatial-learning abilities are superior to those of nontechnical students at two-year colleges in the United States.**

Recently, I participated in an investigation of individual differences in two-year college students' cognitive ability (with particular focus on multiple representations of spatial ability), personality, and academic performance in mathematics.[xii] We separated two-year college students into two groups: *technological* and *nontechnological* students. Technological students were students in programs that frequently receive support from the NSF's Advanced Technological Education program (Robotics, Lasers/Optics/Photonics, and Mechanical Technology). Nontechnological students were enrolled in commensurate workforce programs at the same two-year colleges, such as Early Childhood Education or Criminal Justice. All students in the study (N = 304) were within one to two semesters of completing their

AAS degree, thus the likelihood that they were satisfied with their career choice was higher.

We used a statistical procedure referred to as latent class modeling to determine how many subgroups of two-year college students would be resolved if we allowed them to be clustered based only on their scores on the myriad tests we administered (as mentioned above). This procedure revealed that there were only two classes of profiles in the sample. When we then looked at how many of the students in each class were from technical or nontechnical programs, we discovered that in the first subgroup, 92 percent were from technological education programs. It seems quite clear that students selected to participate in technological education based on their differing abilities.

So how can information from this research be used to distinguish among those students who have a high interest and learning ability to be successful in technical education programs?

- Many students in middle school and early grades of high school who are not necessarily in the top 20 percent of academic achievers should be encouraged to prepare for career-focused postsecondary programs of study below the baccalaureate level in community and technical colleges.
- Content and relevant applications of high school courses for these students should be tailored to more closely meet their interests and applied learning abilities.
- Students with superior spatial learning abilities should be identified and encouraged to enter technician education.

---

[i] Charles A. Murray, *Real Education: Four Simple Truths for Bringing America's Schools Back to Reality* (New York: Three Rivers Press, 2008).

[ii] Ibid., 96.

[iii] Sandra Scarr, "American Child Care Today," *American Psychologist* 53 (1998), 95–108.

[iv] Heather J. Turner, A. Alexander Beaujean, and Darrell M. Hull, "Classification of Students in Two-Year Colleges: A Latent Class Model Approach" (paper presented at the Twelfth Annual Conference of the International Society for Intelligence Research, Cyprus, December 2011).

[v] Olesya Blazhenkova, Michael Becker, and Maria Kozhevnikov, "Object-Spatial Imagery and Verbal Cognitive Styles in Children and Adolescents: Developmental Trajectories in Relation to Ability," *Learning and Individual Differences* 21 (2011), 281-287, doi: 10.1016/j.indf.2010.11.012

[vi] Harold Pashler, Mark McDaniel, Doug Rohrer, and Robert Bjork, "Learning Styles Concepts and Evidence," *Psychological Science in the Public Interest* 9 (2008), 105-119, doi: 10.1111/j.1539-6053.2009.01038.x.

[vii] Ibid.

[viii] Carol L. Gohm, Lloyd G. Humphreys, and Grace Yao, "Underachievement Among Spatially Gifted Students," *American Educational Research Journal* 35 (1998), 515 531, doi: 10.3102/00028312035003515; Lloyd G. Humphreys, David Lubinski, and Grace Yao, "Utility of Predicting Group Membership and the Role of Spatial Visualization in Becoming an Engineer, Physical Scientist, or Artist," *Journal of Applied Psychology* 78 (1993), 250-261, doi: 10.1037/0021-9010.78.2.250; David F. Lohman, "Spatial Abilities as Traits, Processes, and Knowledge," in *Advances in the Psychology of Human Intelligence,* vol. 4., ed. Robert J. Sternberg, (Hillsdale, NJ: Lawrence Erlbaum, 1998), 181-248; David F. Lohman, "Spatial Ability," in *Encyclopedia of Intelligence,* vol. 2, ed. Robert J. Sternberg (New York: Macmillan, 1994), 1000–1007; David F. Lohman, "Spatially Gifted, Verbally Inconvenienced," in *Talent Development: Vol. 2. Proceedings from the 1993 Henry B. and Jocelyn Wallace National Research Symposium on Talent Development,* ed. Nicholas Colangelo, Susan G. Assouline, and Deann L. Ambroson (Dayton, OH: Ohio Psychology Press, 1994), 251–264; Ian Macfarlane Smith, *Spatial Ability: Its Educational and Social Significance (*London: University of London Press, 1964; Jonathan Wai, David Lubinski, and Camilla P. Benbow, "Creativity and Occupational Accomplishments among Intellectually Precocious Youth: An Age 13 to Age 33 Longitudinal Study," *Journal of Educational Psychology* 94 (2005), 785–794.

[ix] Wai, Lubinski, and Benbow, "Creativity and Occupational Accomplishments."

[x] Murray R. Barrick and Michael K. Mount, "The Big Five Personality Dimensions and Job Performance: A Meta-Analysis," *Personnel Psychology* 44 (1991), 1-26; John R. P. French, Jr., Robert D. Caplan, and R. Van Harrison, *The Mechanisms of Job Stress and Strain* (London: Wiley, 1982); Carmi Schooler, "The Intellectual Effects of the Demands of the Work Environment," in *Environmental Effects on Cognitive Abilities,* ed. Robert J. Sternberg and Elena L. Grigorenko (Mahwah, NJ: Erlbaum, 2001), 363–380.

[xi] Cameron Anderson, Sandra E. Spataro, and Francis J. Flynn, "Personality and Organizational Culture as Determinants of Influence," *Journal of Applied Psychology* 93 (2008), 702-710; Gregory J. Feist, *The Psychology of Science and the Origins of the Scientific Mind* (New Haven, CT: Yale University Press, 2006); Gregory J. Feist, "A Meta-Analysis of the Impact of Personality on Scientific and Artistic Creativity," *Personality and Social Psychological Review* 2 (1998), 290-309.

[xii] Turner, Beaujean, and Hull, "Classification of students."

Helping Students Make Career Choices:
The Role of Secondary School Counselors

Pat Schwallie-Giddis, PhD, Associate Professor & Department Chair
Counseling & Human Development, George Washington University

Secondary school counselors are loaded with a variety of tasks to help students, their parents, and schools, and are regularly encouraged to "look to the future." None of their tasks are more important than the guidance they provide for students—and their parents—in helping students make career choices, develop an educational plan, and prepare for the next educational experience after high school graduation. The following advice is provided for counselors who will be assisting in the facilitation of the proposed STEM Technician Program.

Facilitating Communication Between Secondary and Postsecondary Institutions

The success of Career Pathways programs depends on productive interaction between the secondary and postsecondary levels. Secondary counselors have an important role to play in making that interaction a reality, especially when programs are in the planning stages. For instance:

- Secondary counselors should provide postsecondary counselors with the names of Career Pathways students in their schools.
- Secondary counselors should review any changes that postsecondary institutions make in requirements for specialty fields to determine whether the new requirements call for revision of the corresponding secondary curricula.
- Secondary counselors should call one or two meetings annually to discuss procedures, problems, and changes that affect postsecondary enrollment and completion.

Communicating with Area Employers

- Counselors should survey area employers annually to *identify recent graduates who have been employed.*

- Counselors should *keep employers informed about STEM Career Pathway programs* through personal contact, newsletters, and/or electronic communication.
- They should also encourage and work with employers in providing mentors. This involves not only setting up mentoring relationships but also facilitating training opportunities for new employees who would like to become mentors.
- Counselors should take a leading role in arranging for field trips. They should work with teachers in providing preparation and follow-up field trips.
- Counselors should *seek opportunities for employers to interact with STEM Career Pathway teachers*.

Communicating with Community College Counselors
- Secondary counselors should keep in regular contact with their counterparts at area community colleges. They should determine which institutions their students are most likely to attend after graduation and then make sure a counselor or administrator from each of those institutions is invited to serve on the career guidance team, the curriculum development committee, or both. Lack of interaction can cause indifference and even hostility.

Bridging the Gap Between Academic and Technical Teachers
- STEM Career Pathways provide career contexts for learning academic subjects. In practical terms, this means that the teaching of technical and academic content must be integrated. For most teachers, both technical and academic, this is a stretch. Many schools, even by their physical layouts, create environments in which technical and academic courses are thought of as completely separate spheres of activity. Counselors must take the initiative in changing the preconceptions of both academic and technical teachers by encouraging interdisciplinary communication and collaboration.

Disseminating Information about Career Pathways

- Counselors should take the lead in providing information to parents, employers, and the media regarding the STEM Career Pathways opportunities.
- Secondary counselors should also try to ensure that their schools have career centers, either freestanding or as designated areas within their schools' media centers, where students can obtain brochures and other printed items that contain employment information. The centers should provide assessment tools and, if possible, computerized career guidance systems to help students identify their interests, aptitudes, and skills, as well as the careers in which those skills are most useful and in demand. Parents should also be made aware of the materials and services provided by the career center.

Dispelling Negative Misconceptions about STEM Technician Education

One of the greatest frustrations experienced by innovative educators is that parents, students, counselors, principals, academic teachers, and even many employers have a negative image of technician education. Despite irrefutable evidence to the contrary, most people still think that the only way to succeed in today's world is to get at least a bachelor's degree. Many people consider an alternate STEM pathway as a "dumping ground" for students who lack the intelligence to succeed in the engineering and science STEM path. This prejudice may discourage students from giving serious, realistic thought to how they plan to make their living as adults.

This negative image represents a battle that counselors must be willing to fight, no matter how unfavorable the odds might seem. Bettina Lankard Brown has offered the following excellent strategies for combating the negative image of technician education.[i]

- Give students something to brag about.
- Bring parents on board: Address misconceptions about the need for all students to seek college degrees. Describe the STEM option that might better meet the needs of their children. (People who hold technician education in low esteem usually don't know what the real options are for graduates with good

technical skills. They don't know where the good jobs are, what they pay, and what it takes to get them.)

- Target marketing to those who have the greatest impact on student choices: student organizations and business and industry representatives. (Many businesses are more than willing to send representatives to speak with students and their parents.)
- Work with the media.
- Every counselor should read Getting Real: Helping Teens Find Their Future by Kenneth Gray (SAGE Publications, 1999) and Other Ways to Win: Creating Alternatives for High School Graduates by Kenneth Gray and Edwin Herr (SAGE Publications, 2000).

Providing Student Advisement

- Get parents more deeply involved in their children's career planning.
- Organizing events for joint parent-student planning (e.g., Saturday seminars and workshops). One interesting and encouraging outcome of this process is that students can see that the path to success invariably involves overcoming obstacles. Students should also see that even the most successful adults talk about what they would have done differently if, as youngsters, they had known what they know now. When well planned and executed, these events will be very popular among parents and students alike, and will produce excellent results.
- Help students and parents make preliminary pathway decisions—without "tracking."
- Career-awareness activities should begin as early as the elementary grades and continue into the middle school years. By the eighth grade, students should be exposed to the differences between pursuing a career and merely getting a job. During the eighth grade, students should be given both interest and aptitude assessments to help them begin to make initial career choices. It is extremely important that the results of these assessments be explained to both the parents and the

students so that they can see that this is just the beginning of the exploration phase; it is merely a place to start. No decision made in eighth grade, or at any time in high school, should be seen as a "track" that students can't change. The key to the success of this process is to be sure that a student's career plan (which begins in eighth grade) is reviewed each year by the student, the parent, and a school representative (ideally, the counselor).

Helping Students Take the Next Step After High School Graduation

- By the middle or end of the eleventh grade, students should be developing a clear understanding of the knowledge and skill requirement of the career specialties within their pathways. The counselor should make that information available through interviews with employers, written materials, the Internet, guest speakers, field trips, and (where possible) summer employment.

- At the beginning of the twelfth grade, the counselor should conduct one-on-one or group sessions with students to help them review their career plans and set postgraduation goals. Then, if necessary, the student's plan should be revised to reflect where he or she is on the road to reaching personal goals.

- Students should be encouraged to identify barriers that might hinder them from taking the next step, such as finances or whether the student is prepared academically to pass the college entrance exam. If students learn by the eleventh grade that they are deficient in certain academic areas, they still have time to rectify the situation so that they will not require remediation in college.

Encouraging Dual Enrollment and Students' Decisions Regarding Higher Education

- Counselors should encourage STEM technician students to take advantage of dual-enrollment opportunities. Dual enrollment allows students to get a taste of what it takes to do college-level work, it reduces the time required to complete college programs, and it can reduce the cost of college attendance and give students a "can do" attitude about their college programs.

"Learning to Learn" and Career Counseling

- Our country's economy has undergone many dramatic changes in the last few years, and these changes have resulted in vastly different requirements for employment. "Job security" is a thing of the past. Everyone should be prepared for the inevitability of change. Postsecondary education is vital for almost all adults to succeed in careers. And AAS degree programs in STEM fields are the best choice for many students.

Student Portfolios in the Guidance and Placement Process

- Counselors should encourage students to keep their portfolios up to date and in good order. A student portfolio is a record of a student's progress in achieving the goals associated with his or her Career Pathway. Information in the portfolio should include assessment scores, lists of courses taken, grades received, and descriptions and documentation of occupational experiences (including work-based learning or other special training).[ii]

Work-Based Learning

- STEM technician students should have an opportunity to learn about their chosen careers firsthand, before joining the workforce full time as an adult. Through work-based learning experiences, the student learns the importance of teamwork, effective communication, workplace ethics, and problem solving. Through work-based learning experiences, the student also learns whether he or she has the skills, interest and temperament for a particular career. Work-based experiences can greatly enhance a student's understanding of career requirements, both mental and physical. Counselors should strive to ensure that work-based learning experiences have clear objectives and that students understand the objectives and are assessed on their success in meeting them.

In conclusion, it is important for students to understand that even if they do choose a technical career initially, that may lead them into other careers in the future. Many students prepare themselves in a particular field, become successful, and later move into managerial positions. This may require additional education and on-the-job training. In today's world, it is not unusual for a person to have five to seven different careers over a lifetime. The important message is to encourage students to find something that is of interest to them and get them started on a career path that will lead them into the world of work.

---

[i] Bettina Lankard Brown, "The Image of Career and Technical Education," *ACVE Practice Application Brief* no. 25 (2003), http://www.cete.org/acve/docgen.asp?tbl=pab&ID-115 (accessed August 2005).

[ii] Dan Hull, *Career Pathways: Education with a Purpose* (Waco, TX: CORD, 2005).

A HIGH SCHOOL MODEL

High School Institutes of Study:
A Comprehensive Approach to College and Career Readiness

Dr. Jill Siler, former Executive Director of Academic and Organizational Development, Lake Travis Independent School District; current Superintendent of Gunter Independent School District.

Understanding the Context

In the early 2000s, Lake Travis Independent School District (west of Austin, Texas) underwent a severe financial crisis due to a change in state funding, which resulted in drastic cuts, specifically to elective and CTE courses. By mid-2004, the situation had improved, the growth in student population had grown rapidly and the district was ready to rebuild. Instead of just adding back the same programs that were in place before the cuts, the district seized an opportunity to create what *high school could look like through the lens of college and career readiness.*

At the same time, district leadership determined that one large, comprehensive high school was preferred to two smaller ones because it would allow students more diverse course offerings and would save money through streamlined programs and staffing. However, the district was also committed to maintaining our "small" learning environment and began to think about how smaller communities within the context of a large high school might benefit our students. Ultimately, we redesigned our high school to prepare our students for careers and postsecondary study. We did this by creating aligned programs, by creating a *culture around institutes,* and by developing relationships with industry and postsecondary partners.

Creating a Comprehensive Strategy for College and Career Readiness

As we thought about what this comprehensive program would entail, we decided that it should prepare *all students for their next step,* whether that was a four-year university, community college, technical school, or direct entry into the workforce. We developed an approach in which the first year of high school was an exploratory year. We would encourage sophomores to choose which career pathway to explore; they would then spend the next three years deeply pursuing this area of study.

To explore which career pathways to focus, we used student interest data, workforce data and parent and community input. We coupled student interest with regional workforce data to flesh out our programming, and we took our draft frameworks to several community forums for further input. It was important to us that we not just rebuild departments, but rather create interconnected pathways with multiple entry points and exit points that all students could access. Our pathways became the Institutes of Study and included these six areas:

- Institute of Advanced Science and Medicine
- Institute of Veterinary and Agricultural Science
- Institute of Business, Finance, and Marketing
- Institute of Math, Engineering, and Architecture
- Institute of Humanities, Technology, and Communication
- Institute of Fine Arts

Each institute began with a common foundation course of study that included the courses all students needed for graduation. Next was a career exploration emphasis that included all the coherent sequences available within each career pathway. The next layer was postsecondary preparation, which included work-based learning programs; advanced-placement, dual-credit, and Tech Prep course offerings; certifications and licensures; and internships, job shadowing, and site visits. This framework provided students with a clear path on how to best progress through coursework to achieve their ultimate goals.

While students were encouraged to choose an area of study and pursue that in depth, they were also allowed to switch into a different area the following year if they did not like the pathway they were in. Students did not enroll in or graduate from these institutes; instead, the structures served to guide them into a path that could best prepare them for postgraduate success. One way we ensured this was by molding our academic and CTE programs together into single institutes. Our math, engineering, and architecture institute showed our students the rich connections between geometry and engineering design, and our advanced science and medicine would later be coupled with our health sciences program. Our counselors met with each student individually every year from eighth through twelfth grade; they also ensured that the student's core coursework supported the Institute of Study that they were working within.

Creating a Culture Around the Institutes of Study

Because the district was experiencing rapid growth, we were required to expand the high school campus. In our visioning on how to best take advantage of this opportunity, we considered how it might affect students to design the entire high school around the Institutes of Study, so that as students are walking to their math class, they see the engineering design lab and are subtly reminded of the connection across the subjects. This *wayfinding* philosophy included signage throughout the building as well as color-coding throughout the campus to differentiate one institute from another. Other opportunities were taken to bring the institutes to life, such as the Internet café near the technology institute, a java bar run by the business students, and a glass front school store fronting the cafeteria, run by the marketing students.

Developing Relationships with Industry and Post-Secondary Partners

Some our most critical accomplishments were the relationships we created with industry and postsecondary partners. An advisory board oversaw and guided several of the career-based institutes. This advisory board was composed of students, teachers, campus and district leaders, parents, community members, industry partners, and postsecondary educators. This broad base of advice helped staff members shape curriculum choices, acquire articulation agreements with area colleges, and develop job-shadowing and internship opportunities that enabled students to see the connections between what they were doing in high school and their future careers. Our purpose was framed by these three questions:

- Where is it that we want our graduates to be?
- How can we help them reach that goal?
- What will be needed to move this vision forward?

Our programs are grounded in the philosophy that the skills necessary for success in college are the same skills needed for success in the work force, and that every student deserves to exit our school college- and career-ready. Our vision with the Institutes of Study is to create a venue for all students to achieve their goals, regardless of what paths they took after high school. Our institute-based framework followed that vision.

The STEM Technology Future at Lake Travis High School

The Lake Travis High School (LTHS) Institute for Math, Engineering, and Architecture is well grounded in the Project Lead the Way Curriculum and offers five full years of engineering coursework. Recently, LTHS was named one of the ten National Model Schools that offers Project Lead the Way Engineering.

Like most STEM high school programs, the majority of students who complete the entire PLTW program have been high academic achievers who are planning to enter universities for BS studies in science and engineering. However, we know that for every engineer needed in the field, multiple technicians are also needed. We have recognized that the Career Pathways for STEM Technicians initiative that Dan Hull describes in chapter 4 is an important opportunity to more successfully prepare our students for life after LTHS.

Creating Multiple Pathways within the High School STEM Curriculum

One of the main reasons we chose the Project Lead the Way curriculum as the cornerstone of our engineering program was that it was not an elite program. It was accessible to all students. And PLTW provides exceptional professional development to equip teachers to be successful with all students. The first few years of that coursework are instrumental for all students, regardless of what future STEM path they take. Yet, the courses that LTHS offers for upperclassmen are very specialized and often geared toward students who plan to pursue a four-year engineering degree after high school. What we need is a parallel program for upperclassmen that builds off of the foundational PLTW courses and specifically equips students for the STEM technician positions that industries are so eager to fill.

Overcoming the Challenges in Building a STEM Technician Pathway

Several challenges exist in this journey to build a STEM pathway at the high school level that will equip future technicians. These include the availability of quality certifications and licensures, valuable articulation agreements, and relevant college and career counseling.

1. **Certifications and licensures**: We consistently try to build in opportunities for students to gain valuable certifications and licensures in each of our Institutes. While some career pathways have multiple options for certifications and licensures, we're missing quality certifications and licensures that are geared for STEM Technicians.

2. **Articulation agreements**: As more and more high school students are looking for college credit opportunities in high school, it is critical for STEM to enter the arena of articulation agreements. LTHS currently offers thirty hours of articulated credit in the area of Business, Finance, and Marketing. In Engineering and Architecture, that number is fourteen, with only seven credit hours focused in engineering. As students see those credits appear on their college transcript, they realize that it is easier and cheaper to continue down the road of business than it would be to start over in engineering. These articulated credits also need to be in the form of an aligned STEM-technician program that will truly prepare students for the field while they are in high school.

3. **College and career counseling**: This piece is probably the most important of the three. Students need to make fiscally conscious decisions about postsecondary study in light of theeconomic context of today. This includes a realistic understanding of the costs of college as well as the financial gains of degrees at all levels. This also includes knowing the employability of various degree programs and their corresponding entry-, mid-, and exit-level salaries. Finally, while we don't know the careers of tomorrow, we do know where the critical markets are emerging, and that heavily involves STEM education. The demands on today's counselors are heavy; huge workloads often include testing, parent

conferences, and other administrative paperwork. We need to truly rethink the role of today's counselor and build a structure that will better guide students into meaningful career pathways.

LTHS Institutes of Study: Lessons Learned

- Alignment within your programs is key.
  - From your *foundational course of study* to your *career emphasis* to your *postsecondary options*, build an aligned program.
- Think BIG, start small.
  - Even if you are only starting with one program or the entry course in several programs, flesh out what could be, and build a plan to get from where you are to where you want to be.
- Student interest is important but not the only thing that matters.
  - When developing academies, institutes, pathways, etc., don't rely solely on student interest. Instead, gather further data from the workforce commissions, community leaders, and the school community.
- Develop your philosophy.
  - Determine whether students change from one program to another.
  - Decide whether students will apply to join an academy or if they are open to all students.
- Look for stellar professional development.
  - Ensure that your programs are supported by ongoing, rigorous professional development and that they will be sustainable if your current teachers leave.
- Reconceptualize the norm.
  - Consider whether there is a way to structure your building to give deeper meaning to students about the relationship of their core courses to their CTE courses—to build a culture around your pathways.
- Partnerships are critical!

- o Provide meaningful work for your advisory boards and partners so that they truly influence your programs.
- o Broaden your partnerships such that all levels of the industry are represented.
- Get the students out.
  - o Find ways to build in job shadowing, mentoring, fieldwork, internships, college visits, etc.
- Build a program with multiple exit points.
  - o Ensure that your programs don't culminate in an elite program, but rather offer a multitude of opportunities for students at all levels.

Creating Successful CPST Partnerships:
Leadership Roles and Challenges

Dan Hull

Educational programs at two-year colleges to prepare technicians for advanced technology careers are mostly under-enrolled and are not providing the numbers of completers that U.S. employers need if we are to maintain and advance our nation's security and economic superiority. Chapters 4 and 5 propose a model and suggest dual-credit strategies to create an efficient, robust high school pipeline. These chapters call on STEM high schools to provide a program of study to recruit and prepare capable, applied learners who could enter and be successful in AAS-degree technology programs that lead to rewarding technician careers. Chapters 6–14 present examples of advanced technological education programs in selected cutting-edge technologies. These chapters provide information about the various fields and recommend curriculum content and course sequences that will prepare students to become technicians in high-demand occupations. Chapters 6–14 contain models that can be adopted and adapted to meet specific community needs. These chapters are written by the directors of selected NSF/ATE National and Regional Centers of Excellence, and these directors are able, willing, and eager to provide information, planning assistance, and professional development for education/employer partnerships. The last chapter is a "how-to" manual that will enable community and state leaders to plan together to make an urgently needed improvement in education and workforce development.

Communities need to form partnerships of educators and employers to provide the vision and substantiate the need for STEM career pathways. These leaders will need to design the pathways, encourage and recruit students, cooperate to deliver a secondary-postsecondary sequence of courses, and provide the resources necessary to support and sustain the initiative. This work requires leaders and decision makers from three vital groups: secondary schools, community/technical colleges, and employers. The following sections and work charts describe the leadership tasks and roles for each of these three partners.

Secondary-School Leadership Tasks and Roles

1. Commit to providing encouragement and appropriate teaching strategies to a variety of student learners and achievers to help them successfully pursue STEM careers.

This task requires the secondary school to recognize, encourage, and support *all students*—not just the highest academic achievers—as candidates for STEM careers. Some students will choose to become scientists, some will be engineers, and some will be technicians.

Most high schools are conscientiously recruiting and supporting students who are among the highest academic performers in math and science to enter STEM-career preparation—and that is commendable. But many secondary students who are not among the highest academic achievers are capable of learning science, engineering, and technology, and these students may have interests and learning styles that will enable them to learn and contribute effectively as technicians in technical work teams. Supporting and educating these students will require a different curriculum path, different courses, different learning environments, and different teaching styles. It will also require board members, administrators, counselors, teachers, and parents to adopt a different mindset about successful STEM education. It requires a change in what we value. But it also opens unbelievably good career opportunities to a group of deserving students who are often neglected. The budget costs for making this commitment don't have to be excessive—particularly if dual-credit technical courses are introduced.

2. *Value, support, and encourage graduates who seek admission to technician-education programs in community colleges.*

Community and technical colleges offer high-quality education and training at affordable prices. Many more students are now considering the two-year college as a stepping-stone to a four-year baccalaureate degree. But students who pursue an Associates of Applied Science degree can gain the knowledge, skills, and credentials necessary to enter rewarding jobs after the equivalent of two years at college—which doubles the savings. And AAS degrees can lead to very rewarding careers. But they do not have to be terminal degrees. Although many technicians with only AAS degrees enjoy long, productive careers with upward mobility, others choose to continue their education to open new doors and explore new career opportunities. If AAS-degree technicians want to continue their studies in the technical field, they can enroll in an additional two years of education to earn a bachelor of engineering technology (BET). They can also complement their technical background with additional work to earn a four-year (bachelor's) degree in business or education majors. *A community-college technical-education degree does not close the door to additional higher education.*

3. *Provide information and experiences to help students and parents make decisions about pursuing education that will lead to a career as a STEM technician.*

Counselors are the key to achieving this task, but they cannot accomplish it without the help of teachers and employers. This task really needs to begin in middle school—or perhaps even in earlier grades. And the groups that may need the most attention and understanding are the students' parents. Parents, as well as students, need to see firsthand from employers how technicians are employed, how they are rewarded, and how they advance in their careers.

One false perception is that technicians are working in blue-collar or craft jobs. But as I explained in chapter 1, technicians are an essential element of the "three-legged stool" that describes our technical workforce: scientists, engineers, and technicians. The technicians on the work teams where I served understood the science, the math, and the technology, but they uniquely knew how to solve certain problems because they were the "geniuses of the labs"; they could make things work.

Traditional science and math courses turn many students off from these subjects—even in the elementary grades. There are two primary reasons for this: (1) the material may be taught in an abstract manner or without labs that show students how what they're learning can be used, or (2) the students do not achieve at a sufficiently high level to give them confidence in the subject. If the school district embraces a STEM career pathway leading to technician education, teachers should be prepared to tailor their teaching of science concepts to these learners by using real-world applications. Projects, summer camps, and practical labs can also help.

But even when average achievers discover that they really are interested in science and technology, they often don't realize that there are technician jobs available and attainable in which they can have fun with science and technology and also earn a good living. Secondary schools need to invite role models from industry to tell this story.

4. *Develop and initiate an approved STEM curriculum path for eleventh- and twelfth-grade students interested in STEM postsecondary technician education.*

This is a major task that will need to be accomplished in cooperation with two-year colleges and employers. It will also need to be aligned with state and district curriculum standards. NSF/ATE centers can provide models, planning assistance, and professional development for teachers and counselors. I offer more information and suggestions about curriculum development in the second section of this chapter, "Community College Leadership Tasks and Roles."

5. *Cooperate with community colleges to provide academic and technical dual-credit courses.*

There may be some question as to which institution (college or school district) will take the lead on this initiative. Different states also have different policies regarding compensation for high schools and colleges and the transferability of the postsecondary credits that students earn. Chapter 5 provides much more information on dual credit.

6. *Provide teachers with curriculum, materials, supplies, technology, and professional development to effectively teach applied learning to STEM technician students.*

Teachers really need to be a part of the decision and development process. Facilitating academic and technical teacher collaboration should also be a high priority. As mentioned earlier, NAS/ATE national and regional centers are a resource for materials and professional development. See http://www.atecenters.org.

Secondary-school leaders should consider using the following work chart during their decision-making and planning processes.

Leadership Roles: Who Will Take Responsibility?

SECONDARY SCHOOL WORK CHART

TASK	LEADER OR DECISION MAKER	ANTICIPATED ISSUES
1. Commit to providing encouragement and appropriate teaching strategies to a variety of students learners and achievers to help them successfully pursue STEM careers.	1.	1.
2. Value, support, and encourage graduates who seek admission to technician-education programs in community colleges.	2.	2.
3. Provide information and experiences to help students and parents make decisions about pursuing education that will lead to a career as a STEM technician.	3.	3.
4. Develop and initiate an approved STEM curriculum path for eleventh- and twelfth-grade students interested in STEM postsecondary technician education.	4.	4.
5. Cooperate with community colleges to provide academic and technical dual-credit courses.	5.	5.
6. Provide teachers with materials and professional development to effectively teach applied learners.	6.	6.

Community College Leadership Tasks and Roles

1. Work with employers to identify and define STEM-technician career opportunities.

Colleges have assembled a core of employers in particular fields of interest and have engaged these employers in defining the knowledge and skills that they require of technicians. Frequently, this information is called *skill standards*. This database of employment information is then translated into a set of essential knowledge and skills that form the job competencies that define colleges' technical curriculum and course content. The content of the first year of college technical courses in turn defines the knowledge and skills required of high school graduates. Math skills are frequently a defining element that determines the success of incoming technician students. Colleges also poll employers to assess the projected demand for new technicians needed in particular technical fields.

2. Create AAS-degree curricula for STEM-technician career preparation in high-demand areas.

College faculty members design the curricula, courses, and laboratories necessary to prepare students to meet the skill standards. The college curricula then determine the competencies and knowledge that high school applicants need to be successful in their postsecondary technical studies. Frequently, dual-credit courses in technical areas serve as the articulation requirements for successful transition from secondary to postsecondary technical studies. NSF/ATE centers are valuable resources to assist colleges in this task. They have developed national skill standards, a curriculum model, and supporting teaching materials for particular technologies, and colleges can easily adapt these materials to local or regional employer needs.

3. Encourage high schools to consider providing an alternate eleventh- and twelfth-grade curriculum path that will allow

students to successfully articulate into AAS-degree technician education programs.

There are two aspects of this task: (a) creating an interest, and (b) making the decision that this curriculum opportunity will be put into practice. The *interest* aspect depends on high school teachers, curriculum coordinators, and administrators. *Decisions* for adoption will be made by the high school principal, the district superintendent, and the local and state school boards. Decision makers should consider the following information and questions:

- Most rewarding careers require some form of postsecondary education.
- Judging from their postgraduation experiences, how well is the college presently serving applied learners?
- What college and career opportunities do these students presently face? How will providing these pathways enhance their opportunities?
- How can the college and the high school cooperate to make implementation of this decision cost-effective?

4. ***Help high school teachers and curriculum developers design appropriate courses and sequences for eleventh- and twelfth-grade STEM technician curricula. Cooperate with STEM high schools to create and provide academic and technical dual-credit courses.***

Employer-based skill standards and state education requirements provide the basis for design of the AAS-degree curriculum. In turn, the first year of coursework at academic and technical colleges, plus state high school graduation requirements, provide the basis for the design of the eleventh- and twelfth-grade STEM career-pathway curriculum.

College technical faculty can help teachers and curriculum coordinators understand the technologies, design appropriate laboratories, and select equipment. Colleges can teach dual-credit courses to provide initial instructor expertise and to share laboratories. College faculty members can also provide high school teachers with examples of technical problems that they can use to provide an applied context for high school academic courses.

5. *Invite secondary students, parents, teachers, counselors, administrators, and board members to visit the college and review STEM-technician programs and laboratories. Invite employers to attend and describe relevant career opportunities.*

Many of the common misconceptions about technician careers can be alleviated if members of the high school community can visit colleges, tour their laboratories, and listen to faculty presentations. Students, teachers, parents, counselors, administrators, and board members can all benefit from these firsthand experiences.

Where it is appropriate, testimonies from working technicians who have completed the AAS degree can have an enormous impact on people's perceptions of technician careers. In addition, colleges can take advantage of the alumni profiles and video interviews of former students that several NSF/ATE centers have created.

6. *Host technically oriented summer workshops and camps for secondary students, teachers, and counselors.*

Once an interest has been created, colleges can provide more of the knowledge and experience that students need to make career choices by offering summer camps or workshops. These camps and workshops should focus on lab experience and use the same equipment that is used in college labs and in technical workplaces. These camps and workshops are useful for high school teachers and counselors, as well as for students. In addition to hands-on lab experiences, summer camps include presentations and Q&A sessions with employers and former students who are now working as technicians. These technology camps frequently plan field trips for campers to visit employer sites.

NSF/ATE centers such as OP-TEC (http://www.op-tec.org) have prepared monographs that describe successful summer camps, including daily schedules, demonstrations, activities, and helpful suggestions.

College leaders should consider using the following work chart during their decision-making and planning processes.

Leadership Roles: Who Will Take The Responsibility?

COMMUNITY COLLEGE WORK CHART

TASK	LEADER OR DECISION MAKER	ANTICIPATED ISSUES
1. Work with employers to identify and define STEM-technician career opportunities.	1.	1.
2. Create AAS-degree curricula for STEM-technician career preparation in high-demand areas.	2.	2.
3. Create an AAS-degree curriculum for STEM-technician career preparation in high-demand areas.	3.	3.
4. Help high school teachers and curriculum developers design appropriate courses and sequences for eleventh- and twelfth-grade STEM technician curricula. Cooperate with STEM high schools to create and provide academic and technical dual-credit courses.	4.	4.
5 Invite secondary students, parents, teachers, counselors, administrators, and board members to visit the college and review STEM-technician programs and laboratories. Invite employers to attend and describe relevant career opportunities.	5.	5.
6. Host technically oriented summer workshops and camps for secondary students, teachers, and counselors.	6.	6.

283

Employer Leadership Tasks and Roles

1. *Provide information for STEM-technician skill standards and employment projections.*

In addition to the students, employers are the "customers" who receive the benefits of a secondary-postsecondary STEM-technician education program. They are companies who make or use technological products and services, research-and-development laboratories that generate new and emerging applications, and government or military organizations that do a variety of work relating to research, development, application, and product control. Lists of employers within a geographical region who use a particular STEM technology can be obtained from state business databases, professional societies, nearby universities, and trade journals. Frequently, NSF/ATE centers also have employer information.

Employers are also the source of the knowledge and skill specifications that they require of the technicians that they employ. These specifications are typically developed into skill standards that describe what the employers want their newly hired technicians to know and be able to do. Skill standards are usually broken down into *critical work functions, tasks, and technical and employability skills.*

Most NSF/ATE centers have developed or acquired skill standards and have used them to design and develop model curricula. These centers are frequently available to assist colleges in designing appropriate curricula.

2. *Advise in the development of secondary and postsecondary STEM-technician curriculum and course content. Periodically review course content for relevance and currency.*

Colleges also need a representative panel of employers to guide them in the development of their AAS-degree STEM-technician curriculum. The ideal procedure is to start with national skill standards and a national curriculum model, if available. The college program advisory panel reviews the national skill standards and determines how to make them appropriate for employers in the region. Based on this information, the college designs a curriculum and course descriptions to address the revised standards as well as institutional and state curriculum requirements. The curriculum and course content are then submitted to the employer advisory panel for review and comment. If possible, representatives from the STEM high school attend program advisory panels as non-voting members.

In many cases, colleges have already begun assembling and meeting with an employer group to design and update the AAS-degree curriculum before the STEM high school becomes involved. High school CPST leaders should also be invited to meet with these employer groups.

3. *Advocate for CPST programs to colleges, STEM high school administrators and teachers, and policy makers. Encourage and support program development and growth.*

Frequently, a community requires a catalyst to encourage a community college–high school partnership to form and pursue a CPST program. Employers are usually the ideal catalyst, because they need these new technicians and can only support economic development within the community if they have enough well-prepared technicians to carry out their mission.

To produce well-qualified technicians, educational programs must be created and sustained by *secondary-postsecondary partnerships*—employers should insist on this point. Attempts to use high school graduates as technicians are short-sighted: they may produce a cheaper labor market, but these workers will not have the technical knowledge and strength that can be developed through strong scientific and math-based AAS-degree programs.

And students who attempt to gain employment as technicians after high school graduation will likely experience limited job security and poor opportunities for career advancement. Continuing their education after employment will be very difficult, because they need to master fundamental math and science courses before they can attempt postsecondary technical courses.

Likewise, community college technician programs that are not cooperating with local high schools typically struggle to enroll enough students. They also experience inordinate drop-out rates because most of their entering students are poorly prepared in math and science. Faculty from these colleges should avoid the temptation to lower standards in an attempt to realize more course completers and reduce attrition. Employers recognize these kinds of compromises because the new technicians they hire are frequently incapable of performing and advancing in their jobs.

Instead, employers must support and encourage college faculty and administrators to use CPST programs to build a high school pipeline that will produce capable college students.

4. *Host meetings of students, parents, counselors, and teachers at employer sites where STEM technicians can be observed working.*

Many high school students, parents, teachers, counselors, and administrators have a poor perception of technician careers. Employers can provide a more realistic vision of technician careers by allowing these decision makers to see the work and the workplace where technicians are employed.

Employers of technicians should coordinate with high school education leaders to allow employers to host meetings or luncheons for these decision makers at the work site. Employers can very effectively dispel poor impressions of technical careers by explaining technician jobs and career paths, providing tours of the workplace environment, and encouraging educators to interview technicians and their supervisors.

5. *Provide mentoring and internships for secondary and postsecondary STEM-technician students.*

Most secondary STEM-technician students are trying to gain confidence that they have chosen the appropriate career field. Mentoring, job shadowing, and short-term internships can provide a clearer understanding of how the student's chosen field will play out in the workplace. In a postsecondary AAS-degree program students have only two short years to determine the type of job they would like to pursue upon graduation.

Most STEM-technician career opportunities are available in several specialization areas. In addition, careers may be available in research and development, product development, operations, and field services. Internships or part-time jobs for postsecondary technician students provide useful information and experiences to help the student make appropriate decisions when seeking a job after graduation. And job experiences provide an opportunity for students to learn the "soft skills" they need and to learn how these skills will benefit them in their career.

6. *Provide talks, demonstrations, and tours for middle school students to learn and become interested in STEM technician careers.*

Many students lose interest in technical careers between the fourth and eighth grades because they do not have opportunities to learn how the math and science they are studying are used in technology and in their possible careers. Most elementary school students are naturally curious about science, but this curiosity frequently disappears in the absence of some stimulus. Scientific and technical professionals can enhance the science courses in elementary and middle schools by providing talks, demonstrations, and tours of their workplaces.

In middle school, students should begin to explore career options in broad fields, such as science and technology, medicine and health care, business and finance, social services, etc. Working professionals can provide valuable insights to aid students' decision making—or at least help students begin to consider how they might shape their future education and career plans around their interests and aptitudes.

Employers should consider using the following work chart during their decision-making and planning processes.

Leadership Roles: Who Will Take The Responsibility?

EMPLOYER WORK CHART

TASK		LEADER OR DECISION MAKER	ANTICIPATED ISSUES
1.	Provide information for STEM-technician skill standards and employment projections.	1.	1.
2.	Advise in the development of secondary and postsecondary STEM-technician curriculum and course content. Periodically review course content for relevance and currency.	2.	2.
3.	Advocate for CPST programs to colleges, STEM high school administrators and teachers, and policy makers. Encourage and support program development and growth.	3.	3.
4.	Host meetings of students, parents, counselors, and teachers at employer sites where STEM technicians can be observed working.	4.	4.
5.	Provide mentoring and internships for secondary and postsecondary STEM-technician students.	5.	5.
6.	Provides talks, demonstrations, and tours for middle school students to learn and become interested in STEM technician careers.	6.	6.

Concluding Remarks

This is not a book about "trendy ideas" in education; nor is it a report on recent research in the field. Rather, it is a "call to arms" about a significant problem that is affecting our country's ability to be economically competitive and to adequately defend our shores. This problem is also limiting the opportunity for many students to pursue education and career preparation that is interesting, meaningful, and rewarding to them.

Mostly, this book provides models, information, and proven strategies for recognizing and solving the problem. The information and suggestions presented in the preceding chapters come from the experience of educational leaders and practitioners.

This chapter provides recommendations and worksheets to help leaders begin to work on changes proposed in the book. Good leaders will find ways to use and improve upon these resources. I hope that the information, models, and strategies presented in this book will help leaders create educational changes that will contribute to our country's great qualities and the potential of its people.

CPSIA information can be obtained at www.ICGtesting.com
Printed in the USA
BVOW011524091112

305144BV00007B/34/P

9 780985 800611